THE LIBRARY
ST. MARY'S COLLEGE OF MARYLAND
ST. MARY'S CITY, MARYLAND 20686

Hermann Remmert

Arctic Animal Ecology

With 156 Figures

Springer-Verlag
Berlin Heidelberg New York 1980

Professor Dr. Hermann Remmert
Fachbereich Biologie
der Universität
Lahnberge, Karl-von-Frisch-Straße
3550 Marburg/Lahn, FRG

Translated by:
Joy Wieser
Madleinweg 19
6064 Rum/Austria

ISBN 3-540-10169-1 Springer-Verlag Berlin Heidelberg New York
ISBN 0-387-10169-1 Springer-Verlag New York Heidelberg Berlin

Library of Congress Cataloging in Publication Data. Remmert, Hermann, Arctic animal ecology. Bibliography: p. Includes index. 1. Animal ecology-Arctic regions. I. Title. QL105.R4513   591.5'2621   80-18707.

This work is subject to copyright. All rights are reserved, whether the whole or part of the material is concerned, specifically those of translation, reprinting, re-use of illustrations, broadcasting, reproduction by photocopying machine or similar means, and storage in data banks.

Under § 54 of the German Copyright Law where copies are made for other than private use, a fee is payable to the publisher, the amount of the fee to be determined by agreement with the publisher.

© by Springer-Verlag Berlin Heidelberg 1980
Printed in Germany.

The use of registered names, trademarks, etc. in this publication does not imply, even in the absence of a specific statement, that such names are exempt from the relevant protective laws and regulations and therefore free for general use.

Cover design: W. Eisenschink, Heidelberg.

Typesetting, printing, and binding: Brühlsche Universitätsdruckerei, Giessen.

2131/3130–543210

Preface

A large number of comprehensive publications has been devoted to the Antarctic, to its plant and animal life. It is therefore relatively easy to familiarize oneself with the current state of Antarctic research. Nothing comparable is available for the Arctic. The heterogeneity and richness of the northern polar regions seem to have discouraged any attempt at a synthethic approach.

This book has evolved from an attempt to summarize the results of 15 years of ecological and physiological research work in the Arctic – mostly on Spitsbergen. The necessity of comparing our results and the ecological conditions of Spitsbergen with other arctic regions grew into a full-sized book on arctic animal ecology.

It is not meant as an exhaustive survey of the relevant literature. Instead I have tried to show how closely the various fields of research are interwoven, how many questions can be solved if only notice is taken of fellow scientists and their results, and how much arctic animals have in common.

This book would not have been possible without the helpfulness of many colleagues. Above all I should like to mention Professor Rönning and Professor Solem of Trondheim University (Norway), Professor Arnthor Gardasson of Reykjavik University (Iceland), Dr. Nettleship, Dr. Oliver and Dr. Ryder of Canada and Professor West of Fairbanks University (Alaska, USA).

The first two drafts of the manuscript were typed by Dagmar Weidinger-Messner, who also drew the new figures. Mrs. Köhler and Mrs. Riediger gave valuable assistance in the further preparation of the manuscript, which Mrs. Joy Wieser has translated into English. Dr. Konrad F. Springer and Dr. D. Czeschlik encouraged me throughout. My warm thanks to all of them!

Marburg, July 1980   Hermann Remmert

Contents

I. Introduction: Delimitation of the Arctic 1

II. Ecological Factors in the Arctic 7
    1. The Diurnal Rhythm of Organisms and Its Relationship to Environmental Parameters 7
    2. Temperature Conditions in the Arctic 24
    3. Ecological Factors Other Than Temperature and Light 58
    4. The Combination of Factors in the Arctic 65

III. (Almost) Common Characteristics of Arctic Animals . . 70
    1. Population Cycles 70
    2. Seasonal Migrations of Birds and Mammals . . . 82
    3. Entrainment of Animals to the Yearly Cycle . . . 93
    4. The Importance of the Proximity of the Ocean: Marine Birds and Mammals 103
    5. The Ratio of Productivity to Biomass in the Arctic 107
    6. Species Problems 109

IV. Peculiarities of the Systems 122
    1. General Principles 122
    2. Stability and Constancy of Arctic Ecosystems . . . 129
    3. The Animals in Terrestrial Ecosystems 131
    4. The Animals in Limnic Ecosystems 137
    5. The Animals in Marine Ecosystems 141

V. Types of Arctic Climates 155

VI. Case Studies 161
    1. "Warm" Arctic: A Section Through Northern Scandinavia from Tromsø (Norway) to Kevo (Finland) 161
    2. Arctic Alaska: Comparison Between Prudhoe Bay and Point Barrow 176
    3. High-Arctic Continental: the Canadian Archipelago 183
    4. High-Arctic Oceanic: Svalbard or Spitsbergen . . . 190

5. Arctic Lakes 203
6. The Arctic Seas of the Old World 214
7. Antarctica: a Comparison 225

References 235

Subject Index 247

I. Introduction: Delimitation of the Arctic

Whereas the antarctic continent itself is identical with the antarctic terrestrial ecosystem, it is a far more difficult matter to define the limits of the Arctic. Indeed, it often seems that mere naive impressions or the special field of research of the investigators concerned have been responsible for the variety of definitions found in the literature. Botanists and physical geographers have frequently taken the forest limit as representing the boundary of the Arctic, whereas others have preferred to make use of the length of the growing season. Climatologists have employed annual mean temperatures, as expressed for example by the 0 °C annual isotherm or by the 10 °C isotherm for July. Social geographers have claimed that the Arctic begins at the northernmost limit of agriculture, thus associating it with special forms of human culture. Soil scientists and engineers have defined the Arctic on the basis of the distribution of permanently frozen ground, with all of its attendant problems (structured soils, pingos, construction problems). Zoologists and botanists have postulated connections between the distribution of characteristic plant and animal species and the limits of the Arctic. Depending upon their personal preference oceanographers have defined the High Arctic or the Arctic as the region consistently covered with ice in summer or in winter (Figs. 1, 2).

Although the boundaries often run parallel with one another, or are even congruent, they may equally well show considerable divergence, and discrepancies of a thousand kilometres are not uncommon.

In practice not one of these lines provides a truly satisfactory delimitation of the Arctic. The circumpolar forest limit is made up in some places of birch (Betula), in others of spruce (Picea) and pine (Pinus) and in some places of larch (Larix). The determining factors differ considerably from region to region. Along the coasts, for example, the forest limits are displaced far to the south, not as a result of the temperature conditions, but on account of the strong winds prevalent. Man himself has assisted in the process by clear-felling: since the large quantities of water in very rainy regions cannot be utilized by the few remaining trees, moors develop which in turn prevent recolonization by tree species. Examples of this can be seen on the west coasts of Scotland and Ireland. Both of these regions were included in the "Tundra" project of the IBP programme and comparisons were made with the "Tundra" regions of Norway, Sweden, and the Canadian Archipelago. Warm oceanic currents push the ice limits far to the north along the coasts of Norway and Iceland, so that the harbour Longyearbyen on the west coast of Spitsbergen remains free of ice in some winters despite its latitude of 78 °N. On the other hand, polar oceanic currents moving in a southerly direction push the ice limits far south in some places, such as on the east coast of Greenland. The length of the growing season is highly dependent upon winter temperatures, i.e., whether the winter is warm and oceanic or cold and continental, but the vegetation in continental regions is often much

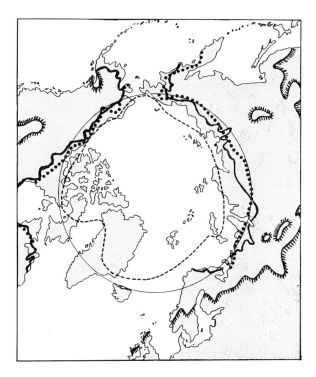

Fig. 1. Limits of the arctic. Forest limit ———; limit of cultivation ▨▨▨; Arctic limit according to Nordenskiöld ·····; southern limit of the High Arctic – – – –; Arctic Circle ——

richer than in oceanic regions with equally short growing seasons. In fact, melons can be cultivated successfully in continental regions almost up to the permafrost, whereas in oceanic regions like Europe the limit of melon cultivation is several thousand kilometres further south. Not all of the permafrost nowadays in existence is the product of today's climate. Relics of permanently frozen ground from glacial times exist in Switzerland and can be localized by the retarded growth of the trees above them. As to the use of temperature in defining the Arctic, it obviously has its drawbacks. The 0 °C annual mean isotherm or the 10 °C July isotherm, for example, both run through tropical mountains as well as the Arctic and give no indication of the vast differences between the two regions. And as to the northernmost limit at which agriculture is practised, this is displaced far to the south along the coasts. At the same time we know of primitive communities on the equator that support themselves by hunting and fishing and practise no form of agriculture, although the reasons are very different from those applying in the Arctic.

Quite clearly, the criteria discussed in the foregoing are not merely naive, subjective and dependent upon the particular field of interest of the investigator concerned: not one of them can be employed universally, and not one of them has been strictly worked out. Where the incongruencies between the postulated limits and the true Arctic become too apparent, help is sought in constructions such as the Nordenskiöld line (Fig. 1).

Nevertheless, there is a clear mathematical boundary to the Arctic and this is the Arctic Circle, which girdles the regions where in summer the sun does not set for at

Introduction: Delimitation of the Arctic

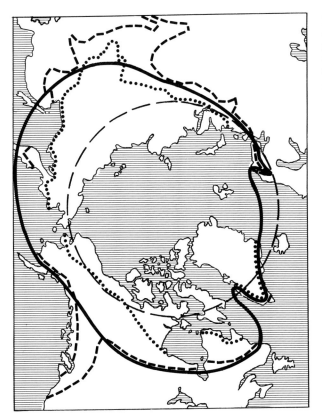

Fig. 2. Limits of the Arctic. Continuous permafrost ———; southern limits of the discontinuous permafrost – – – –, 0 °C annual isotherm ·····; Arctic Circle — — —

least 24 h, and where in winter complete darkness prevails for at least 24 h. Although it has often been included as an additional piece of mathematical information in the literature it has never been employed as a definition of the Arctic. At first sight it might appear that a definition of this kind has neither geographical nor ecological significance. Nevertheless, the midnight sun has always held a remarkable fascination for Man, and the question which we shall now try to answer is whether this mathematical boundary is as irrelevant as a cursory glance would suggest. This is quite definitely not the case. Troll long ago drew attention to the fact that the arctic climate is a purely seasonal one and undergoes no daily fluctuations, i.e., a situation exactly the opposite of that prevailing in tropical mountain regions where the climate is purely diurnal (Troll, 1955; Figs. 3–6). On Kilimanjaro, Mount Kenya or in the tropical Andes the same temperature can be measured at any one time of day all the year round, although from one time of day to another it differs considerably. In the Arctic, on the other hand, summer and winter are quite different, but the temperatures typical of each of these seasons, the summer with its perpetual daylight and the winter with perpetual darkness, remain constant throughout. Now the activity of poikilothermic organisms, such as the bacteria and fungi which play an essential role in soil formation, is exponentially dependent

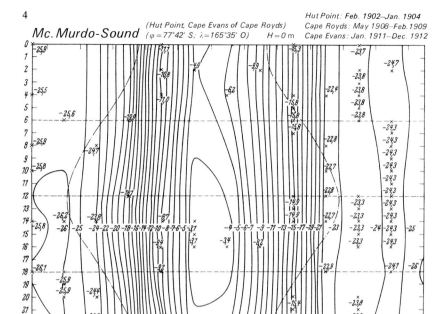

Fig. 3. Thermoisopleth diagram of a station with an antarctic coastal climate. (The *space* between the *dot-dash lines* indicates the length of the polar days and nights). *Ordinate:* time of day. (Troll, 1955)

upon temperature. It is therefore of enormous significance whether temperatures are relatively constant, especially in the lower ranges of the usual optimum for biochemical processes, or whether they fluctuate about a mean value and are thus sufficient to accelerate the biological processes in poikilotherms for a few hours daily. This explains why mean temperatures, even those of the growing season, as indicated by the 10 °C isotherm for example, convey little information of any value. Exactly the same applies to solar radiation per day and unit area. The vital point is whether this radiation is fairly evenly distributed over the 24 h, as it is in the Arctic, or whether it is confined to about half of the day, in which case the hourly values would have to be twice as high. Consequently, daily or monthly mean values for incoming radiation convey very little biological information. Another factor is the internal clock in plants and animals which, as we now know, is indispensable to the organism. This clock completes a revolution once in approximately 24 h. However, it has to be "reset" regularly by external factors, its zeitgeber usually being the alternation of light and darkness. What happens in the uninterrupted darkness of the arctic winter or in the continuous daylight of its summer? And lastly there is the fact that the ratio of the length of daylight to length of darkness, or the "photoperiod", changes over the course of the year and is the factor responsible for the ability of plants and animals to coordinate their activities with the annual march of seasons. Is this mechanism still effective north of the Arctic Circle?

Evidently the biological implications of a mathematical delimitation of the Arctic are greater than at first sight apparent. We shall therefore make use of the

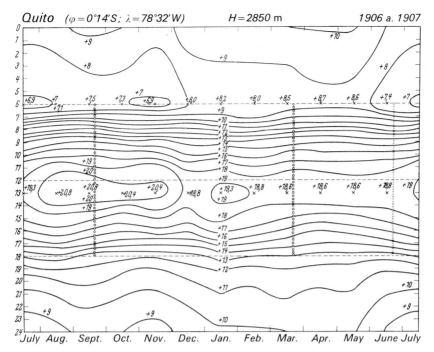

Fig. 4. Thermoisopleth diagram of the equatorial highland station at Quito. Beneath, seasonal distribution of precipitation. *Vertical lines* (··· or - - -) indicate the zenithal and lowest positions of the sun. (Troll, 1955)

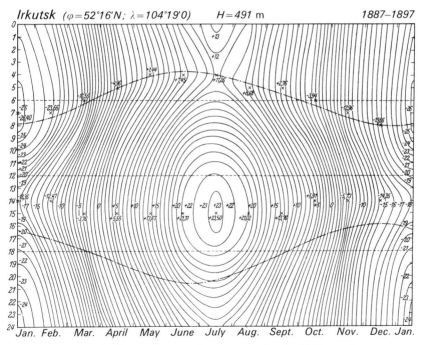

Fig. 5. Thermoisopleth diagram for Irkutsk as an example of a highly continental station of cool temperate latitudes. (Troll, 1955)

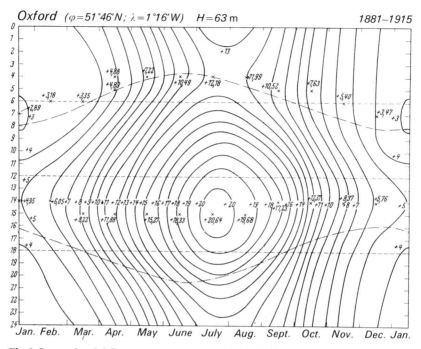

Fig. 6. Seasonal and daily temperature variations in Oxford, depicted in a thermoisopleth diagram. (*dot-dash lines* indicate the times of sunrise and sunset). (Troll, 1955)

Arctic Circle as providing an objective criterion with which to support the claim that we are dealing with conditions that are typically arctic, accepting the fact that it includes a number of different types of landscape and excludes others that are often considered to be arctic.

Unfortunately, such a strict definition cannot be adhered to throughout. In earlier studies on the ecology of arctic animals attention was seldom paid to the position of the site under investigation relative to the Arctic Circle: a wealth of valuable observations in fact originates from points further south. The transition from a seasonal and diurnal type of climate south of the Arctic Circle to a purely seasonal climate north of it is not abrupt but takes place gradually. In regions immediately south of the Arctic Circle there is little difference between the quantity of radiation entering during the day and that entering at night, so that there is very little difference between daytime and nighttime temperatures. To the immediate north of the Arctic Circle there are 24 h of continuous daylight only once during the whole summer. Some investigations carried out slightly south of the Arctic Circle have been included although they do not strictly belong here.

In accordance with our definition we have excluded most of the southern third of Greenland as well as the larger part of the Canadian mainland including Hudson Bay. Only a narrow strip of northern Alaska is considered. Of Siberia only a small northerly strip, so narrow as to be hardly recognizable in the European part of the Soviet Union, is included. On the other hand, a very substantial portion of the Scandinavian peninsula falls within our definition, although Iceland does not. In Scandinavia, agriculture is practised on a small scale even north of the Arctic Circle and the forest limit even extends several hundreds of kilometres to its north. Moving eastwards the limits of agriculture and of the forests recede much further south, and only in central Siberia (and in central Canada) does the forest limit push further northwards again.

The course taken by our boundary—the Arctic Circle—is full of surprises and may even appear at first sight to be unrealistic in places, but let us see where it will lead us.

II. Ecological Factors in the Arctic

By choosing the Arctic Circle to define the limits of the Arctic we have automatically brought diurnal rhytmicity and the factors depending upon it into the forefront of our discussion.

1. The Diurnal Rhythm of Organisms and Its Relationship to Environmental Parameters

In the High Arctic, for example on Spitsbergen, the sun remains above the horizon during the whole of the growth season, which roughly covers the months of June and July. From mid-April until mid-August the sun circles the sky. Nearer the Arctic Circle the growth season is often longer than this, even though there are fewer days with midnight sun. Even here, however, there is a continuous "white" light. This phenomenon of apparent continuous illumination during the arctic summer early attracted the attention of scientists. Every reliable investigation on song-birds (Palmgren, 1935; Franz, 1943; Hoffmann, 1959a; Remmert, 1965b; Haarhaus, 1968; Pohl and West, 1973; Krüll, 1976b), however, has clearly confirmed that in all regions north of the Arctic Circle activity is still synchronized with the rotation of the earth. Whether a species or a population of a species is of more southerly or more northerly distribution makes no fundamental difference to its activity. Even when southerly forms are deported to a region north of the Arctic Circle their daily rhythm is strictly synchronized with the rotation of the earth, as could be shown for the greenfinch (Carduelis chloris) on Spitsbergen, and for golden hamsters (Mesocricetus auratus) and Tupajas (Tupaja belangeri) near the Arctic Circle (Daan and Aschoff, 1975; Krüll, 1976). Conversely, northerly forms like the snowy owl (Nyctea scandiaca) or the flying squirrel (Glaucomys volans) exhibited no basic differences in behaviour when moved to southern latitudes (Figs. 7, 8, 9). As would be expected for the theoretical reasons put forward by Aschoff and Wever (see also Daan and Aschoff, 1975), the resting period of light-active song-birds shifts further into the evening with lengthening light period. This holds for the geographical position (resting period during the summer begins earlier in the evening on Spitsbergen than at the Arctic Circle), and for the season (resting period at mid-summer is earlier than in spring or autumn; Fig. 7; Andersson and Müller, 1978). Long-term studies on insects gave identical results. Hatching from the pupae is still linked to the time of day, and the drift in running water is subject to a diurnal rhythm even in the High Arctic. Orientation by means of the sun has been demonstrated for birds (Hoffmann) and spiders (Papi, Remmert) to the north of the

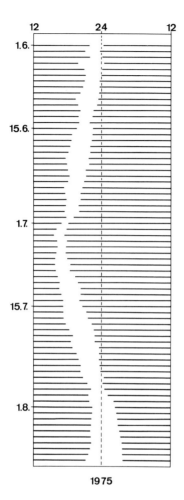

Fig. 7. The phase position of the activity of caged snow-buntings from 1. June to 10. August, 1975, in Abisko, Sweden. (Andersson and Müller, 1978)

Arctic Circle. The daily pattern of raising and lowering of its leaves by the clover plant (Trifolium) is still maintained at a latitude of about 70 °N in northern Scandinavia. The same applies to the opening and closing of flowers (such as Taraxacum and Ranunculus). Deviations from this strict synchronization of rotation of the earth and activity of the organism have only been described on a few occasions. The majority of such reports involved short-term studies employing what would today be dismissed as inadequate methods, and could in some cases immediately be disproved. However, it is not so easy to find an explanation for the rest of these results. In some cases they are due to an annual cycle in the time of day at which maximum activity occurs. In the arctic summer, for example, some small mammals are mainly active during the "dark" hours, whereas in the arctic winter they shift their active period to the lighter hours. A similar situation has been demonstrated in a large number of fish (Fig. 10; Erkinaro, 1970; Müller, 1970a, b). In the final analysis, of course, the activity of such organisms is still strictly correlated with the rotation of the earth, and only occasionally does the activity "escape" its influence for a few days during the changeover from "day active" to

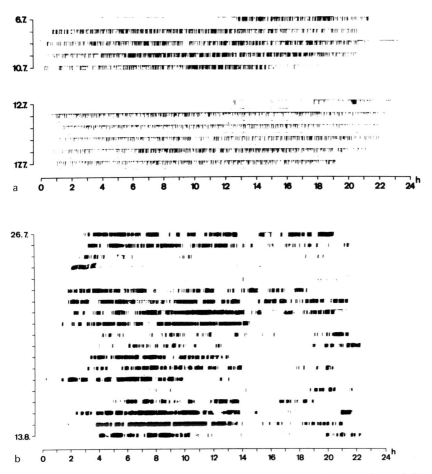

Fig. 8. a Feeding activity of two pairs of snowbuntings in Longyeardalen. A resting period is strictly observed between about 22 h and 1 h (Krüll, 1976b). **b** Perch-hopping activity of a German greenfinch on Spitsbergen. The animal rests between approximately 22 h and 1 h. Its activity is synchronized with the rotation of the earth (Krüll, 1976b)

"night active". It is unjustifiable, however, to postulate a desynchronization of the physiological clock on the mere strength of a "desynchronization" that lasts no more than a couple of days. Even in temperate and warm latitudes similar phenomena occur when an animal shifts its activity to a different time of day, as for example when song-birds shift part of their activity to the dark hours during the migration season. Other examples of apparently continuous activity, implying uncoupling of the physiological clock from the rotation of the earth (if the existence of a physiological clock is not altogether denied in these animals) have been reported in non-passerines. Examples of this kind are of very little value. The overall activity of the organism is by no means an infallible criterion for phase state or for the existence or non-existence of a physiological clock. The reason why non-passerine birds have so far been little studied with respect to rhythm as compared to

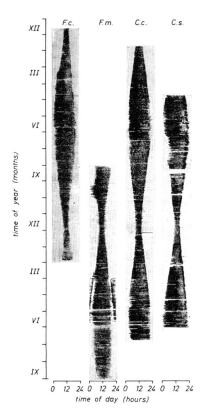

Fig. 9. Recordings of the activity of a chaffinch (*F.c.*), a greenfinch (*C.c.*) and a siskin (*C.s.*) in Messaure (67 N), Sweden. The animals are synchronized with the rotation of the earth throughout the entire year (Daan and Aschoff, 1975)

passerines is that the activity of the latter is so clearly defined and is obviously governed by a physiological clock. Non-passerines, on the other hand, are capable of activity under the most varied conditions and in some cases in temperate and even in tropical latitudes exhibit both daytime and nighttime activity. Although there is no doubt whatever that these birds possess a physiological clock, their activity is still an unsuitable parameter for its detection. The observation of continuous activity in auks (Alcidae), ducks (Anatidae), seagulls and terns (Laridae) is not contradictory to the view that these species are able to synchronize their physiological clock with the rotation of the earth in the High Arctic, even if this has not so far been demonstrated. In none of the descriptions of activities running free of the rotation of the earth is it clear whether this is a truly specific arctic phenomenon. The unequivocal demonstration (Bovet and Oertli, 1974) that the activity rhythm of wild beavers (Castor canadensis) at 51°N in Canada can be free-running suggests that the question should be treated with caution. The same applies to plants in the High Arctic. No daily rhythm in the movements of flowers or leaves could be recorded on Spitsbergen. However, the growth of mosses from Spitsbergen under continuous light in the laboratory was distinctly poorer than under the conditions prevailing on Spitsbergen itself (Kallio and Valanne in Wielgolaski, 1975). This suggest that plants, too, synchronize their physiological clock in the High Arctic with the rotation of the earth.

The Diurnal Rhythm of Organisms

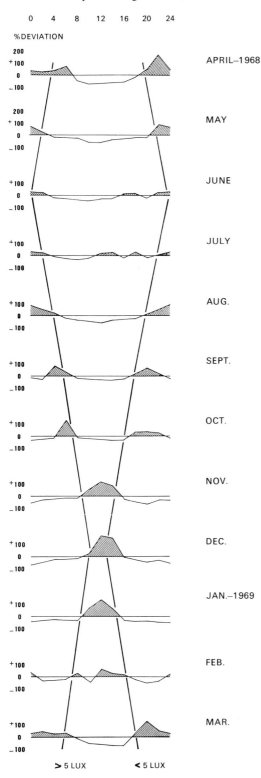

Fig. 10. The activity rhythm of Cottus poccilopus over the course of the year. During the dark winter months the animal shifts its activity to daytime, whereas in summer it is active during the hours of darkness. At all times, activity is synchronized with the rotation of the earth (Müller, 1970a)

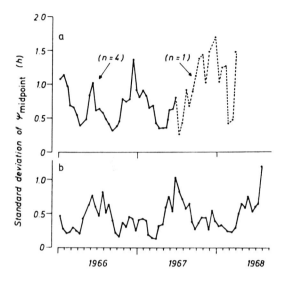

Fig. 11. Standard deviation around the 15-day mean value of phase-angle differences, ψ midpoint, measured between midpoint of activity and **a** midnight (Mesocricetus) or **b** noon (Tupaia, n=3), plotted as a function of time of year. Data computed from continuous records of wheel-running activity at 66°42′N. The standard deviation for the tupaia is conspicuously large during the arctic summer, and that for the dark-active hamster during the arctic winter. (Aschoff et al., 1972)

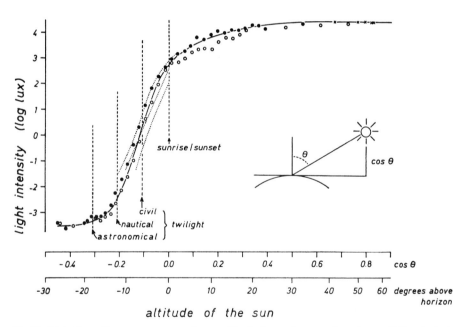

Fig. 12. Light intensity measurements taken at 10-min intervals on a clear day at Erling-Andechs (47°58′, 11°30′E), plotted as a function of the altitude of the sun. *Solid* and *open circles* represent measurements in the evening of August 8, 1972, and in the morning of August 9, 1972, respectively. *Crosses* denote intensities beyond the capacity of the photometer used. The *solid curve* is eye-fitted. *Dashed lines* indicate time of onset of astronomical twilight (sun −18° from horizon), nautical (−12°) and civil twilight (−6°), and sunrise sunset (0°). Of the *three dotted lines*, the *middle one* indicates measurements taken by Scheer (1952, Fig. 2) at 49°52′N, 8°39′E, while the *outer curves* are redrawn from the ranges of intensity measured by Hjorth (1968, Fig. 12) at 57°42′N, 14°28′E. (Daan and Aschoff, 1975)

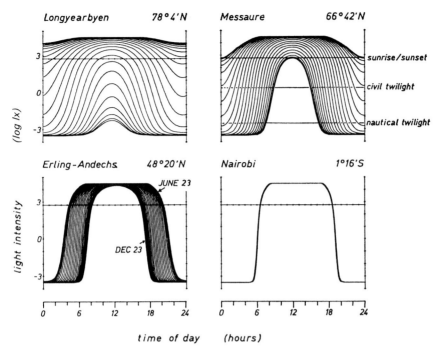

Fig. 13. Calculated curves of light intensity at the zenith under clear sky ("zcs light intensity") at four localities at different latitudes. The uppermost curves always represent June 23, the lowermost curves December 23. The curves between are for semi-monthly midpoints. In the curves for Messaure, the zcs light intensities at sun altitudes of 0°, −6° (beginning or end of civil twilight), and −12° (nautical twilight) are indicated. (Daan and Aschoff, 1975)

What are the zeitgebers responsible for this? Whatever the effective zeitgebers they cannot be very strong, which is suggested by the precision with which the activity of the animals is synchronized with the rotation of the earth (Fig. 11). Long-term recordings made near the Arctic Circle (Messaure, Sweden, 66° 42′ N) revealed a regularly recurring annual cycle in the degree of precision of synchronization, with a minimum during midsummer. Since the temperature over the course of the day is largely dependent upon the path of the sun, temperature fluctuations in the Arctic are quite small (see p. 124) Figs. 4, 5, 22. Further, temperature fluctuation as zeitgeber for the diurnal rhythms in organisms has only been demonstrated in a few cases. In warm-blooded animals temperature fluctuations seem to be unimportant, and even among the poikilotherms there are at least some forms that are equally independent of temperature with regard to the phase relationship between their daily rhythm and their day-night cycle (see e.g. Beck, 1973). We are therefore obliged at this point to consider in more detail the question of light as a zeitgeber. Its significance becomes apparent from figures showing the paths of the sun in the Arctic and the illumination curves calculated from them (Figs. 12 and 13). These figures make it clear that fluctuations in illumination during the course of an arctic summer are very small. Can differences of this magnitude bring about a synchronization of the physiological clock? All laboratory data speak against this

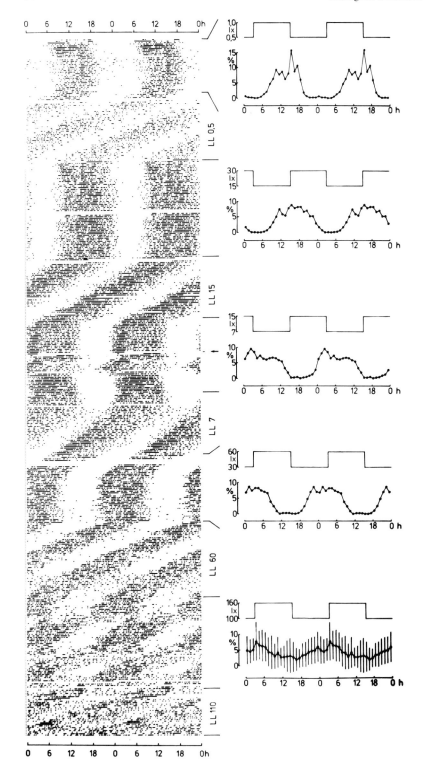

 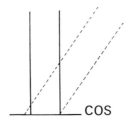

Fig. 15. The recordings of a measuring instrument obey the cosine law. The question is, whether this holds for a free-moving animal. (Remmert, 1976)

possibility (Krüll, 1976a). The stronger the overall light intensity, the larger the difference between light and dark required to exert a zeitgeber effect. Cycles involving an abrupt change from an illumination of 600 lx by day to 300 lx by night never entrain the rhythm of passerines (Fig. 14). This appears to exclude light rhythm itself as a zeitgeber. The situation is made more complicated by the fact that a free-moving bird is not necessarily exposed to the same light intensities as those registered by a fixed recording element. The latter is usually shielded from direct solar radiation in order to rule out errors due to the cosine law (Fig. 15), whereas an organism may very well be exposed to the full effect of the direct sunrays. Animals are in a position to vary the degree of direct radiation reaching them, or even avoid it altogether by means of body movements: they can either fully expose the openings of their recording instruments to the light or they can move them into the shade. This leads to unpredictable errors in the picture they receive of the actual light situation. Light as a zeitgeber can apparently be dismissed. Nevertheless, there are two further parameters to be considered in connection with light: one of them is its spectral composition, which can alter according to the height of the sun, and the other is the azimuth position of the sun which, if taken in combination with landmarks, might also be a possible zeitgeber. Both of these possibilities have been investigated by Krüll (1976a, c).

Exact recordings of the daily changes in the spectral distribution of light in arctic latitudes are not so far available. Some idea can be obtained, however, by registering the colour temperature. Even in the northernmost regions of the Arctic where life is still found, the colour temperature is subject to a regular diurnal rhythm (Krüll, 1976a; Fig. 16). It is effective as a zeitgeber for many passerine species in the laboratory at approximately equal energy inflow and illumination (in lx; Demmelmeyer and Haarhaus, Demmelmeyer, Krüll, 1976b). The sensitivity of the birds to this zeitgeber is enhanced during sexual activity or when high doses of sexual hormone are administered (Demmelmeyer, 1974; Krüll, 1976b). That the same zeitgeber is probably also effective in nature was demonstrated by Krüll in an elegant experiment carried out in the very narrow valley of Longyearbyen, Spitsbergen's largest town. He recorded the activity of greenfinches (Carduelis chloris) brought from Germany, as well as that of snowbuntings (Plectrophenax nivalis) breeding there. The valley only receives direct sunlight around midnight, but for the rest of the 24 h it is entirely in the shadow of the surrounding high mountains. A comparison of the recordings showed that there was no difference in

---

Fig. 14. Attempt to synchronize the activity of a songbird by various strengths of illumination at identical threshold values (1:2) between length of light and darkness. Synchronization is only achieved at a relatively low illumination. (Krüll, 1976c)

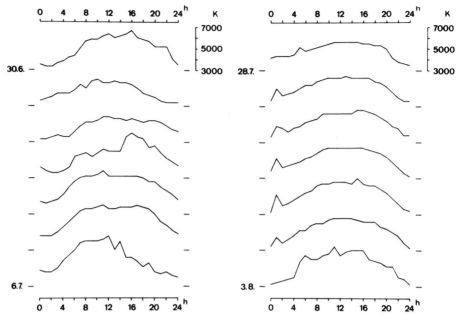

Fig. 16. Daily course of colour temperature for two optional weeks, one from the beginning and one from the end of summer. The *strokes at both ends of each curve* mark 3000 K. (Krüll, 1976b)

the activity of birds in the valley and in the open tundra. They all rested at about midnight, which means that the birds in Longyear valley were asleep during the only part of the day when direct sunshine fell on the valley. Since the lightest hours of the day were passed in sleep it cannot be the light intensity itself that is the zeitgeber.

Such results create more questions than they resolve. It now becomes clear why the use of the unit "lx" in studies concerning diurnal rhythms is justifiably the target of repeated criticism. For an insect whose perception of longwave radiation (red) is poor, a daily pattern in the spectral composition of light may even imply very pronounced subjective fluctuations in strength of illumination over the course of the day, which, however, are not recorded by a photometer calibrated to human sight. The phytochrome system of plants is also probably more readily responsive to colour temperature as a zeitgeber than the human eye or a photometer. Attempts to rear plants of the High Arctic under uniform continuous light must thus be rejected as being unbiological. A thorough analysis of diurnal rhythms in the High Arctic is thus only possible on the basis of an exact knowledge of the daily patterns of variation in the individual portions of the spectrum and a familiarity with the spectral sensitivity of the receptors of the species under investigation.

Even the sun's movement about a fixed point over the 24 h in conjunction with landmarks can have a zeitgeber effect (Fig. 17; Krüll, 1976b). An artificial sun was caused to complete one rotation per 24 h around a cage containing a bird. The result was a strict 24-h rhythm in the experimental animal. If the sun was stopped, however, the bird's own spontaneous frequency took over. A factor of this nature could nevertheless fulfil no more than an auxiliary function in nature since it could

The Diurnal Rhythm of Organisms

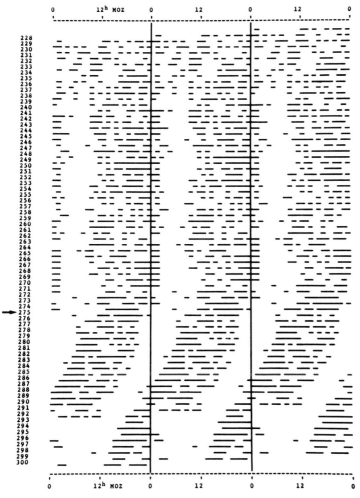

Fig. 17. Activity record for a canary. Beginning on the 275th day (*arrow*) the circling of the "sun" was stopped, but the bulb kept switched on. (Krüll, 1976a)

only exert a zeitgeber effect in animals with good powers of vision and non-roaming habits.

Attempts have repeatedly been made to find other zeitgeber for daily periodicity. In view of the demonstration that at 80°N there is still a regular rhythm in the spectral composition of the light, and that it can function as a zeitgeber in birds, such attempts appear unnecessary. This is especially true in the case of claims that extremely small daily fluctuations in temperature could be adequate (Andersson and Müller, 1978, for the starling in Abisko). All experimental data on birds confirm that unbiologically large and sudden alterations in temperature are

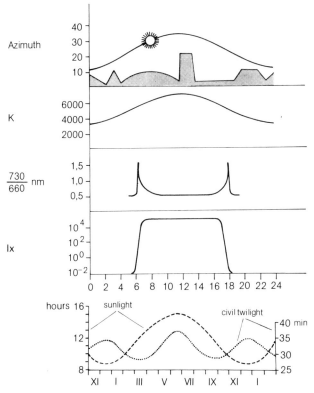

Fig. 18. Changeover from light to dark as a zeitgeber. *From top to bottom* azimuth position of the sun in relation to landmarks in the High Arctic (summer); colour temperature of light in Kelvins in the High Arctic (summer); dark red: light red ratio of light over the course of the day (June, Washington D.C.); light intensity in lx; extremely bright day at about 40°N in June (schematic); theoretical length of the sun's day and the civil twilight at 45°N. (After various authors, from Remmert, 1976)

necessary to synchronize their daily periodicity with a temperature cycle. The small diurnal fluctuations in temperature during mid-summer are certainly inadequate. So far the only investigations providing good evidence that temperature may act as a zeitgeber are those of Kureck (1969) on Simuliidae in the neighbourhood of Messaure (Sweden).

It thus becomes clear that a variety of components are involved in the zeitgeber effect of "light" on organisms. The azimuth position of the sun with respect to landmarks and the spectral composition of light have both clearly been shown to exert a zeitgeber effect in the northernmost regions of the Arctic at which plant and animal life are still found. Despite continuous light arctic organisms are thus able to synchronize their physiological clock with the rotation of the earth. The same zeitgebers can effect synchronization in plants and animals from southern latitudes as well (Fig. 18).

The absence of a daily rhythm in vertical migrations of animal plankton is unlikely to be due to their inability to perceive a diurnal zeitgeber. The explanation is more probably to be found in the characteristic changes in temperature conditions of ocean and lakes with increasing latitude. As a consequence of the very low angle at which the sun's rays hit the surface they bring little heat, so that the surface water is scarcely warmed up at all and a thermocline seldom forms. Thus no metabolic advantage is to be gained by migration to other water layers during the period when the organism is not feeding (McLaren, 1964). The observations of Haney (verbal communication) indicate that planktonic animals such as Daphnia do in fact perceive the diurnal rhythmicity of environmental conditions. Their food uptake is subject to a strict diurnal rhythm even though daily patterns of vertical migrations are no longer detectable.

What do these results tell us? Their only immediate implication is that in the High Arctic, too, the physiological clock can be synchronized with the rotation of the earth. But they tell us nothing about the length of time for which an animal is active or a plant photosynthesizes at these latitudes. We know that light-active animals are theoretically capable of activity throughout the 24 hours and that a plant can carry on photosynthesis around the clock, but does this happen?

Obviously it does not. In discussing the diurnal rhythms of arctic passerines earlier on, we mentioned that they observe a period of rest during the "night", and that this is still observed at a latitude of almost 80°N on Spitsbergen. Laboratory studies indicate that lemmings (Dicrostonyx, Lemmus) also have a distinct diurnal pattern of activity, including a regular resting period. Almost nothing is known of the activity patterns of High-Arctic animals like reindeer (Rangifer), the polar fox (Alopex lagopus) or seals (Pinnipedia). In the Antarctic Müller-Schwarze (1965, 1966) found that the Weddell seal follows a strict pattern of activity with a rest at midday, whereas the activity of the Adelie penguins (Pyoscelis adeliae) in the same regions reaches a maximum in the late morning hours. Neither do the plants seem to carry on photosynthesis uninterruptedly throughout the day.

None of these findings, however, exludes the possibility that under certain circumstances organisms may be active at a time when they normally rest: the external conditions prevailing are conducive to every form of activity. In contrast to the situation in tropical regions where light-active forms are helpless at night, animals of the High Arctic are well capable of avoiding their enemies during the "night" hours, or of attacking them (as regularly happens if a polar fox, Alopex, approaches the sparsely populated colonies of arctic terns, Sterna macrura or paradisea, during their relative resting phase), or of warning of their approach (the snow bunting, Plectrophenax, habitually does this in settlements with cats, or if a polar fox approaches the nesting area). No one has yet followed up the biological and ecological implications of the ability to be active during what is normally the resting phase. It seems probable that the prey-predator relationship is completely different from that seen in tropical regions. It might be speculated that the ability to recognize and scare off enemies during the resting period was one of the important factors in the colonization of these regions, and that the possibility of carrying on photosynthesis throughout the day may have played a role in assisting plants to penetrate regions sometimes exposed to unpredictable conditions.

The primary production of marine planktonic algae at approximately the same temperatures is much higher in the Antarctic than in the Arctic. In drawing a comparison of this kind, nevertheless, it should be remembered that very different geographical latitudes are being compared. In general, the light supplies of antarctic

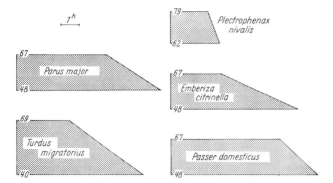

Fig. 19. Prolongation of activity of a light-active bird in northern latitudes as compared with southern latitudes. (Remmert, 1965b)

planktonic algae are much more favourable than those of their arctic counterparts (no continuous darkness in winter). This is a simple consequence of the much lower latitudes of the Antarctic Ocean as compared with the Arctic. Since temperature has less influence on photosynthesis than on breakdown processes, a higher primary production is the result.

A light-active organism in the Arctic is active for much longer periods at a time than one in temperate latitudes (Fig. 19). This can be seen particularly clearly in species with a wide range of distribution, such as the redpoll (Carduelis flammea), which breeds from the High Arctic to (as a different geographical race) the Alps (46°N). It has been suggested that the longer period of activity is responsible for the relatively rapid rate at which the young birds mature in the High Arctic, on the assumption that since the parents have more time in which to bring food to the nest, development might be accelerated. Although at first sight this is an attractive hypothesis, there is little supporting evidence. No comparative counts are as yet available to prove that High-Arctic populations really do feed their young more frequently per 24 h than southern populations. Another point to be borne in mind is that the insects which form the main food of passerine birds are invariably smaller in the High Arctic than in the southern regions (see p. 32). It is therefore questionable whether the portions received by the young birds are as large as in the south. Finally, the adults expend far more energy in catching the smaller prey than farther south, although the caloric content of the High-Arctic insects is higher. It is therefore not surprising that Wagner (1958) found the length of the nestling period of the starling (Sturnus vulgaris), the wheatear (Oenanthe oenanthe), the white wagtail (Motacilla alba) and the meadow pipit (Anthus pratensis) to be no shorter on Röst at 67° 30′ N than in Switzerland. (Belopolski (1957), however, found convincing evidence of exactly the opposite situation in arctic Alcidae, which exhibit a marked reduction in the time spent in breeding and nesting with increasing geographical latitude. It is difficult at present to offer even a speculative explanation for this).

The continuous darkness of the arctic winter represents a complete contrast to the continuous daylight of its summer. However, polar light and the moon, especially if the ground is snow-covered, can produce relatively high intensities of light, and at about midday, notably near the Arctic Circle, it does become lighter for a few hours. Only warm-blooded organisms are capable of activity on land in the Arctic: poikilothermic forms can at the most become active in the ice-free parts of the Arctic Ocean, so that the number available for study is in any case smaller. Synchronization with the rotation of the earth seems to be possible almost everywhere. The Weber-Fechner law states that the perception of differences increases with decreasing strength of stimulus. This means that perception of the differences between nighttime and twilight presents no problems at the prevailing low light intensities; it can therefore be assumed that even in the northernmost inhabited parts of the High Arctic winter-active organisms are able to synchronize their physiological clock with the rotation of the earth. It is a much more difficult matter to answer the question as to which organisms are genuinely active in winter to the north of the Arctic Circle. A number of mammals (cf. p. 91) remain in the same area throughout the winter and cannot avoid the months of continuous darkness. Examples are the reindeer (Rangifer) on Spitsbergen and north Greenland, the muskox (Ovibos) of north Greenland, wolves (Canis) and weasels (Mustela). Certain birds, notably the ptarmigan (Lagopus), are also obliged to spend the winter in such extreme northerly regions.

Some of the views expressed above are more hypothetical than proven. We know of almost no observations as to which forms are really active—a result of the darkness! Reports of unseen polar bears (Thalarctos) roaring a few metres away are legion (cf.e.g., Hilmar Nois in Berset, 1957). Although birds and mammals have often been reported as visitors to winter camps in the Arctic, this tells us relatively little: some were probably individuals that had failed to link up with others of their species for migration to the south and were thus forced to rely upon the proximity of human beings in order to survive. Our data are restricted to a few chance observations, mostly from northern Scandinavia, where several species of birds pass the winter, although it is not clear whether they are entirely independent or only able to exist near human habitations. In every case they were seen near street lamps at Christmas time (winter solstice). It is not known at present where falcons, snowy owls, ravens, ivory gulls and many other species pass the winter. This would be an interesting point for future study. A particularly important question concerns the fact that the energy that can be found and consumed during the brief period of activity with just sufficient illumination, is insufficient to carry a warm-blooded body through the rest of the very cold night. Wallgren (1954) demonstrated this in a comparison between the gold hammer (Emberiza citrinella) and the ortolan bunting (E. hortulana). Nothing is known for certain about the strength of illumination in the arctic winter, nor do we have any reliable facts concerning the optical sensitivity of the animals living there. How can animals like the snowy owl (Nyctaea scandiaca) and the falcon (Falco rusticolus) possibly catch their prey? How can fish-eating birds like Alcidae and cormorants (Phalacrocorax) catch their prey under water where light conditions are even worse? The frequent claim that fleeing swarms of fish reflect light is at best a hypothesis lacking scientific support.

In fact we have no concrete knowledge of the life of animals and birds in the arctic winter, a fact well illustrated by the old discussions concerning the winter sleep of the polar bear. Its young are born during the winter and from what we know about mammals it seems impossible for this to happen whilst the mother is asleep.

In Fairbanks (Alaska, 65°N) redpolls (Carduelis flammea) are active for about 10 h and grouse (Lagopus lagopus) for 5 ½ h at a time when there are only 3 h of daylight. Observations made at 69°N in January showed that grouse feed, fly and utter calls even when it is barely light enough for a human being to see at all (West and Norton).

Thus we are forced to rely largely upon studies and registrations made in open-air enclosures for information about the daily rhythms of animals during the arctic winter. All of these studies have been carried out in the immediate vicinity of the Arctic Circle (Fig. 9). The figure shows that the typical pattern to be expected involves a very long period of activity during the arctic summer and a very short period of activity during the arctic winter in light-active organisms. In the long run the situation for dark-active animals is the reverse. Some species that are dark-active in summer become light-active in winter, for example voles (Microtus spp., Erkinaro) and some fish (like the burbot, Lota lota; Fig. 10). Such observations, however, provide no more than a general idea, since the animals had unlimited food at their disposal. It can be assumed that under the harsher conditions prevailing in their natural environment dark-active forms are confronted with problems during the arctic summer, particularly if they are also hibernators, like the flying squirrel (Glaucomys sabrinus). Conversely, light-active forms have their problems in the arctic winter but, as can be seen from Fig. 10, even light-active forms are confronted with difficulties in the arctic summer and dark-active forms in the winter. The duration of activity is apparently genetically fixed and thus the potential active period is not fully exploited in winter by the dark-active forms nor in summer by those that are light-active. This suggests that light conditions in the Arctic Circle may themselves constitute a biogeographical limit that has so far quite clearly been underestimated.

The enormous differences in the relative length of light and darkness would seem to be eminently suitable as zeitgebers for entraining organisms in the annual sequence of events. Ought long-day plants to be in the majority here (Fig. 20)? Little

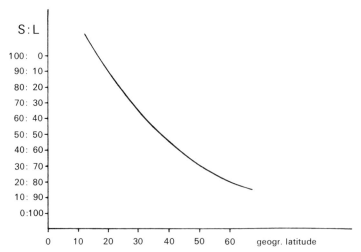

Fig. 20. Proportion of long-day (*L*) to short-day (*S*) plants at various latitudes. (After Remmert, 1965a)

Fig. 21. Growth of tree seedlings of different geographical provenance at different photoperiods; very short photoperiod, *VSD*; short, *SD*; long, *LD*; continuous light, *24*. (Vaartaja, 1959)

is known about this phenomenon, however. The growth of plants of one and the same species (Fig. 21) under various light conditions differs enormously according to the site of origin (e.g., Vaartaja). Longitudinal growth of grass roots from the tundra around Point Barrow, Alaska, is terminated in summer by the decreasing amount of daylight (Shaver and Billings, 1977). Cladocers (Daphnia) from 45°N and 71°N are stimulated by different photoperiods (in combination with other factors) to produce resting eggs. At 45°N their production takes place only when the photoperiod is less than 13 h, but at 71°N when it is less than 22 h (at a temperature of 12°C). Laboratory findings are not entirely in agreement with field data, presumably because of additional factors. Firstly, the photoperiod in water may in some cases differ from that in air (refraction of light in water), and secondly the colour temperature of light (see p. 15) may play a role (Stross, 1969; Stross and Kangas, 1974). Southern populations of the carabid Pterostichus nigrita are very exactly entrained in the annual march of seasons in their environment whereas populations of the same species in Lapland exhibit no photoperiodic induction (Thiele, 1976) and develop at lower temperatures than central European individuals.

From these investigations it seems very probable that photoperiod can play an important role in High-Arctic latitudes. The colour temperature might possibly elicit photoperiodic effects in both plants and animals (induction of dormancy).

2. Temperature Conditions in the Arctic

Although lunar rhythm plays a large role in tropical latitudes, both inland and on the ocean (Corbet and Tjönneland), and is important on the coasts in temperate latitudes (cf. Neumann), its role begins to decrease in northerly temperate latitudes. There is no lunar rhythm at all in the arctic summer since the continuous daylight robs the moonlight of its significance.

2. Temperature Conditions in the Arctic

During the arctic summer, depending upon the geographical latitude, the sun either does not set at all, or merely sinks slightly below the horizon, and then only for a very brief period. As a result, there is scarcely any nocturnal heat loss. Conversely, there is little incoming heat in the arctic winter. Consequently the temperature is very even, both in the growth season and in winter. Diurnal fluctuations in

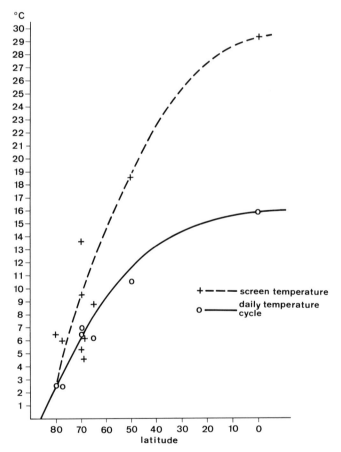

Fig. 22. Magnitude of diurnal temperature fluctuations in July and mean July temperature at various sites at different latitudes. Partly based on short-term investigations and partly on values for only one year

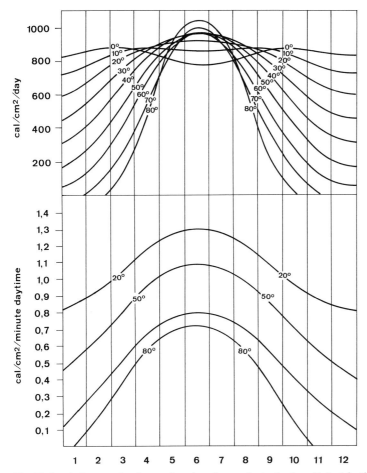

Fig. 23. Insolation per day (*upper figure*) and per minute daytime (*below*) in the course of the year (*abscissa*) at different latitudes. These are calculated values on the basis of the solar constant and height of the sun above the horizon for horizontal areas

temperature lessen towards the pole and increase towards the Arctic Circle and farther south. Recordings (Fig. 22) made at a variety of points in the Arctic have confirmed what was at first merely a theoretical postulate. It often comes as a surprise that the figures for daily incoming energy during the arctic summer are just as high as those for places near to the equator. In the one case, however, the radiation is evenly distributed over the 24 hours, whereas in the other (near the equator) it is concentrated into half of this time (Fig. 23). Accordingly, temperatures rise to a considerable height in equatorial regions but are very even in the Arctic. On the basis of such considerations Troll (1955) employed temperature recordings to develop a system of climate diagrams depicting the different climates of the earth. Thus the climate in the Arctic is characteristically seasonal (Figs. 3–6), whilst that of equatorial mountains is diurnal. Regions between the two extremes occupy an intermediate position. A comparison of the climate diagram of Irkutsk

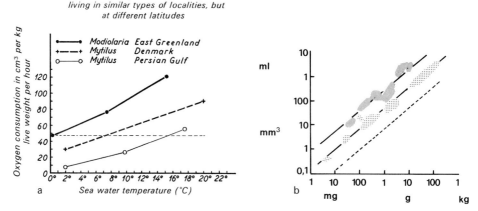

Fig. 24. Oxygen consumption **a** of mytilidae of various provenance at different temperatures. Greenland specimens consume almost as much oxygen at low temperatures as individuals from the Persian Gulf at high temperatures. (After Thorson in Hedgepeth, 1957); **b** of tropical (*shaded areas*) and arctic (*dotted areas*) crustaceans in relation to body size at different temperatures. The oxygen consumption of the arctic specimens is elevated above the calculated line. (After Waterman, 1965)

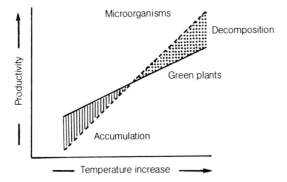

Fig. 25. The influence of temperature on the productivity of green plants and respiratory breakdown processes of animals and micro-organisms. In a hot, wet tropical climate the breakdown of dead material takes place more quickly than the building-up processes (humus layer very thin), whereas in cooler regions breakdown is slower than the reverse process (humus layer is thicker). (After Beck from Remmert, 1978)

(often held to be in the Arctic) with other diagrams clearly shows that it bears no similarity to that of an arctic climate whatever, but a remarkable similarity to that of Oxford. Troll's climate diagrams demonstrate in a striking manner the climatic peculiarities of the regions between the poles and polar circles.

As a rule, the mean daily values for temperature and for incoming radiation are given in publications. In general, however, these values are not biologically relevant since biological processes are exponentially dependent upon temperature. This holds equally for bacteria, fungi and poikilothermic animals, and only the photosynthesis of green plants presents an exception, photochemical processes being largely independent of temperature. The Q_{10} for photosynthesis is generally about 1.5, whereas the Q_{10} for all functions of bacteria, fungi and poikilothermic animals is between 2 and 4 (Fig. 25). Apparently there is no genuine adaption of metabolism, growth and development to low temperature ranges. A relatively high oxygen consumption has been noted in marine crustaceans, molluscs and terrestrial mites of cold areas (Figs. 24 and 25; see also Table I; Block and Young, 1978). Their

Table 1. Oxygen consumption of some arctic invertebrates. (From Procter in Bliss, 1977)

Species	°C	Dry wt mg mean ± SD		Oxygen consumption			
				$\mu l\ O_2\ ind^{-1}\ h^{-1}$ mean ± SD		$\mu l\ O_2\ mg^{-1}\ h^{-1}$ mean ± SD	
Enchytraeidae							
Henlea nasuta	2	0.0371	0.0422	0.0308	0.0319	0.7675	0.2594
	7	0.0310	0.0291	0.0344	0.0326	1.1377	0.6097
	12	0.0336	0.0186	0.0689	0.0417	1.9188	0.5791
Crustacea							
Daphnia pulex	2	0.0430	0.0032	0.0587	0.0186	1.3690	0.4186
	7	0.0441	0.0054	0.1090	0.0193	2.4897	0.4933
	12	0.0432	0.0089	0.2058	0.0248	5.0027	1.4051
Prionocypris	2	0.1239	0.0082	0.0416	0.0074	0.3354	0.0535
glacialis (Sars)	7	0.0755	0.0055	0.0569	0.0150	0.7508	0.1891
	12	0.1125	0.0209	0.1196	0.0214	1.0902	0.2148
Cyclops magnus	2	0.1045	0.0062	0.0410	0.0063	0.3982	0.0628
	7	0.1117	0.0127	0.1480	0.0881	1.3080	0.7390
	12	0.0964	0.0106	0.1497	0.0516	1.5615	0.5483
Cyclops magnus	2	0.0194	0.0022	0.0224	0.0079	1.2026	0.4511
(juvenile)	7	0.0182	0.0015	0.0535	0.0182	2.9549	0.9933
	12	0.0189	0.0042	0.1047	0.0239	5.5785	0.9445
Attheyella	2	0.0102	0.0005	0.0119	0.0058	1.1632	0.5493
nordenskioldii	7	0.0083	0.0011	0.0234	0.0143	2.6927	1.4333
	12	0.0089	0.0008	0.0346	0.0076	3.9691	1.0798
Acarina							
Trichoribates	2	0.0147	0.0007	0.0036	0.0014	0.2436	0.0922
polaris Hammer	7	0.0145	0.0008	0.0052	0.0015	0.3589	0.1159
	12	0.0141	0.0008	0.0060	0.0015	0.4254	0.1060
Hermannia	2	0.0402	0.0038	0.0059	0.0015	0.1469	0.0342
subglabra Berlese	7	0.0438	0.0056	0.0108	0.0047	0.2513	0.1123
	12	0.0451	0.0069	0.0144	0.0061	0.3066	0.1059
Lebertia porosa	2	0.2849	0.0314	0.0446	0.0195	0.1596	0.0735
Thor, S. Lat.	7	0.2670	0.0893	0.2063	0.0755	0.8336	0.4114
	12	0.3157	0.0649	0.3612	0.1305	1.1912	0.4978
Collembola							
Hypogastrura sp.	2	0.0188	0.0034	0.0052	0.0028	0.2732	0.2509
nr trybomi	7	0.0255	0.0081	0.0172	0.0066	0.6953	0.2436
(Schott)	12	0.0166	0.0052	0.0168	0.0102	1.1960	1.0317
Folsomia	2	0.0025	0.0001	0.0003	0.0011	0.1322	0.4766
agrelli Gisen	7	0.0034	0.0004	0.0021	0.0010	0.5708	0.2869
	12	0.0042	0.0006	0.0050	0.0033	1.2164	0.8730
Lepidoptera							
Gynaephora	2	0.2109	0.0610	0.0714	0.0190	0.3536	0.1079
rossi (Curtis)	7	0.1814	0.0091	0.1506	0.0609	0.8319	0.3373
(larvae)	12	0.1837	0.0051	0.4250	0.1098	2.3177	0.5873
Muscidae							
Spilogona prob	2	2.9590	0.1889	1.3737	0.5012	0.4643	0.1666
tundrae Schnabl.	7	3.6520	0.7211	1.7687	0.7232	0.4817	0.1851
(larvae)	12	3.5752	0.4102	2.0021	1.1566	0.5544	0.3124

Table 1 (continued)

Species	°C	Dry wt mg mean±SD		Oxygen consumption			
				$\mu l\ O_2\ ind^{-1}\ h^{-1}$ mean±SD		$\mu l\ O_2\ mg^{-1}\ h^{-1}$ mean±SD	
Chironomidae							
Psectrocladius sp.	2	0.4829	0.0754	0.2966	0.0847	0.6230	0.1854
(larvae)	7	0.5605	0.0915	0.7966	0.4126	1.4126	0.6748
	12	0.4440	0.0321	1.0759	0.4266	2.4010	0.9053
Procladius	2	0.5831	0.0768	0.2834	0.1065	0.4791	0.1396
culiciformis (Linné)	7	0.3965	0.1636	0.3137	0.2212	0.7975	0.4338
(larvae)	12	0.4378	0.2550	0.8280	0.4131	1.9026	0.7194
Cricotopus sp.	2	0.4597	0.0692	0.2798	0.0832	0.6218	0.2061
(larvae)	12	0.4920	0.0409	1.0675	0.6579	2.1999	1.3752
Orthocladius sp.	2	0.1072	0.0203	0.0412	0.0089	0.3900	0.0866
(larvae)	7	0.0785	0.0129	0.0656	0.0304	0.8687	0.4403
	12	0.0982	0.0292	0.1103	0.0510	1.1278	0.5228

respiration rate is almost as high as that of tropical forms at tropical temperatures. The conclusion that such a high oxygen consumption must be reflected in growth and productivity is false: the growth rate of animals of the High Arctic is only very slightly accelerated above the curve obtained by extrapolating from the growth curve of tropical or boreal animals (Krogh's normal curve). We can postulate the following sequence of arguments, although it is still to a large extent hypothetical:

1. The optimal range for biochemical processes is between 27° and 30° C, and here reactions are very fast.
2. Reactions can proceed at lower and at higher temperatures at (almost) the same speed. However, the enzymes involved are different and are very unstable. Their continual rebuilding costs very large amounts of energy.
3. A further possibility would be to raise the concentration of the enzyme, but this is also very costly in terms of energy.
4. Energy (=food) is not available in unlimited quantities, apart from which the very act of finding more food also consumes energy.
5. Therefore the temperature compensation of one process is achieved at the expense of that of another (e.g. growth), and usually at the cost of plasticity with respect to environmental factors (the organisms tolerate only a very narrow temperature range and are highly sensitive to changes in the chemical make-up of their surroundings).

For a summary of the molecular mechanisms involved in temperature compensation see Behrisch in Wieser, 1973; Hoffmann, 1976-78; Hochachka and Somero, 1973. Just how restricted the possibilities are is demonstrated by a comparison of populations of the carabid Pterostichus nigrita (Fig. 26). Although the northern population is able to develop more rapidly at low temperatures than the German populations (Ferenz, 1955) the differences are not very striking. In general it can be said that organisms that can still develop at low temperatures do so,

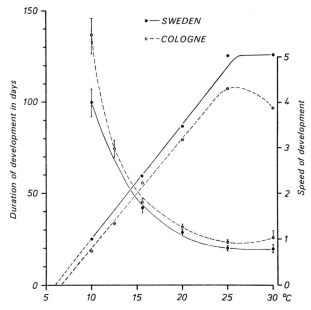

Fig. 26. Duration of development (mean value with standard deviation) and speed of development of central European and north Scandinavian individuals of Pterostichus nigrita at various temperatures. (After Ferenz, 1975)

only extremely slowly (e.g. Teckelmann, 1974). The most spectacular evidence for this is the existence of refrigerators and deep-freezers, which would be useless if micro-organisms had been able to adapt their speed of reproduction to the temperatures prevailing inside them. There would have been time enough in the course of evolution for such micro-organisms to have adapted in the cold regions of the earth, which have been in existence ever since life became possible on our planet. Apparently the absolute limits for the existence of life are reached at such temperatures. The ability to develop at low temperatures is thus always bought at the price of very slow development. If bacteria are unable to develop rapidly at constant low temperatures it is even less likely that higher animals will be able to do so. The following example illustrates the slowness with which organic substances decompose in the High Arctic.

A lone wolf in the Canadian arctic Archipelago killed a six-year-old muskox bull in May 1968, devouring only a small part of it. The wolf returned to feed on the cadaver at long and irregular intervals, and from June to August of the following year (1969) it was accompanied by a she-wolf, both of them then feeding on the remains (Gray, 1970).

Some further examples (see also Table 2):

On Ellesmere Island Collembola require two years to attain sexual maturity and may then live for a further four years (Bliss, 1977); moths may take up to ten years to develop, their body weight increasing only by a factor of 2.2 annually; Mysis relicta from Lake Char in the Canadian Archipelago takes two years to develop, whereas the same species in southern Canada completes its development in one

Table 2. Duration of life cycle of arctic invertebrates. [Bliss, 1975; Sendstad et al., personal observations (Devon Island and Svalbard); Wielgdaski, 1975 (Hardangervidda)]

	High Arctic (seasons)	Hardangervidda	Body length (mm)
Acarina	34	2–4	1
Araneida	8	3	3
Collembola (Hypogastrura tullbergi)	4– 5	1–2	2
Lepidoptera (Gynaeophera groenlandica and G. rodii)	7–11		12–15
Diptera, Nematocera (Sciaridae, Mycethophilidae)	2– 4		4– 6
Hymenoptera symphyta	4[a]		6
Hymenoptera apocrita	2– 3		4– 5
Coleoptera, Carabidae (Amara)		3	6
Gastropoda (Vitrina pellucida)		3–5	4– 5

[a] Probably an underestimate

year; Salvelinus alpinus, the arctic char, takes twelve years to achieve a length of 45 cm in Frobisher Bay of Baffin Island, but may live to an age of more than 24 years (Grainger, 1953; Lasenby and Langford, 1972). The antarctic bottom-dwelling amphipod Bovallia gigantea takes four years near the southern Orkney Islands to reach a length of 4.5 cm, its growth rate being almost constant throughout the year (Thurston in Holdgate, 1970). The same holds for krill (Euphausia spp.). The report that the growth rate of the mollusc Macoma balthica is the same in the southern part of Hudson Bay, Scotland and Netherlands is not surprising when it is considered that this is an intertidal form and that the places in question are all at very similar latitudes. This means that the solar radiation and the degree to which the animals are warmed up in the shallow water are comparable, and temperature measurements made in the open sea cannot be considered as evidence to the contrary (Green, 1973).

Temperatures that fluctuate diurnally around a constant low value result as a rule in a shorter period of development than constant low values. This was predictable from Kaufmann's calculations. These theoretical postulates have since been confirmed by a wealth of evidence (Fig. 27). Organisms that develop at constant low temperatures require more time to develop than those exposed to diurnally fluctuating temperatures, even if the mean values are identical. Accordingly, the productivity of bacteria and poikilothermic animals in equatorial mountain regions is distinctly higher than at the same mean temperature in the Arctic, and larger poikilotherms can develop in a shorter period of time than in the Arctic. Development extending over several years is generally a selective disadvantage (mortality can only be made good by reproduction at intervals of several years, see Fig. 28). This disadvantage plus the brief duration of the growing season in arctic regions ought to result in a gradual increase in the mean size of poikilothermic animals from the High Arctic towards the south, in proportion to the increase in amplitude of the diurnal temperature fluctuations. And this is indeed the case (Fig. 29). (In water the situation is different because the annual period of growth is longer than on land, Fig. 30.)

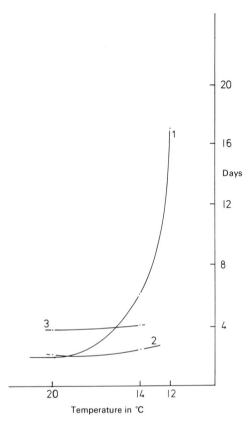

Fig. 27. Time in days between hatching of imagines of Drosophila and the deposition of the first batch of eggs. *1* Constant temperature conditions; *2* diurnally fluctuating temperature conditions; *3* theoretical, calculated curve for alternating temperatures. (After Remmert and Wünderling, 1970)

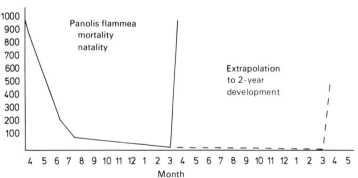

Fig. 28. Of 1000 eggs laid by Panolis flammea only about 10 individuals survive to breed in the following year. If this is extrapolated to a two-year development the species could no longer exist

A further point of significance is the fact that food in the form of plant matter is relatively difficult to break down, and the animals have to break down sufficient food per unit of time to cover their energy needs. With decreasing temperatures (and decreasing amplitude of daily temperature fluctuations) the breakdown of

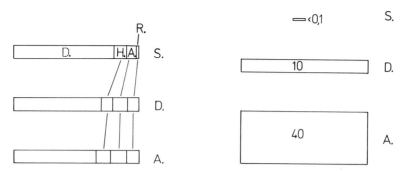

Fig. 29. *Left* distribution of systematic groups in Pitfall traps on Spitsbergen (*S*), on Dovre-Fjell in southern Norway (*D*) and in the Alps (*A*). *D* Diptera; *H* parasitic Hymenoptera; *A* Aranea; *R* remaining groups. *Right* contribution of the remaining groups (*R* in lefthand portion of figure) to the total biomass of the catches in formalin traps on Spitsbergen, in southern Norway and in the Alps. The remaining groups are almost negligible on Spitsbergen, but in the Alps they account for 5% of the individuals and make up 40% of the biomass, i.e., they concist of very large inselcts. (After Remmert, 1980)

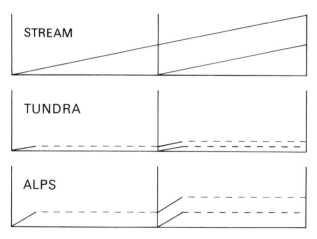

Fig. 30. Schematic representation of the growth of ectothemic animals over the course of two years in a mountain stream in temperate latitudes (*top*), in the tundra (*centre*) and in the Alps (*bottom*) at the same mean temperature. Temperatures are relatively constant in the mountain stream in winter and summer so that the animals are able to grow almost uninterruptedly and attain a fair size within one year. The tundra has constantly low temperatures during its short growing period, whereas in the Alps there are large fluctuations in temperature although the mean value is the same as in the tundra for the equally short growing period. Thus least growth is achieved in the tundra. The conditions in the stream also hold for the ocean. (After Remmert, 1972)

plant food matter becomes less efficient, due to a drop either in the individual's own efficiency or in the activity of symbiontic micro-organisms. The numbers of poikilothermic herbivorous organisms would therefore be expected to decrease from warmer to cooler regions. Schramm (1972) first demonstrated this in model experiments using moths and beetles, and later clearly proved it to be a mechanism in herbivorous reptiles in the subtropics of North America (see also Merkel, 1977).

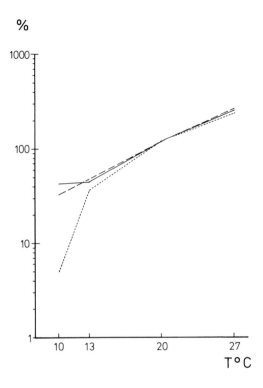

Fig. 31. Growth (in 4) of butterfly larvae (Laphygma) at various temperatures on different foods. --- normal plant food; – – – a synthetic imitation of the above; —— synthetic diet with no indigestible components such as cellulose. (After Schramm, 1972)

We thus have an explanation for the almost complete absence of herbivorous insects in the Arctic and their restriction to continental regions with plenty of sunshine (Fig. 31).

Other effects are probably involved which, however, still lack any physiological explanation. Apparently many poikilothermic animals step up their rate of reproduction under the influence of a diurnal fluctuation in temperature. Evidence for this has been obtained in freshwater rotifers (Brachionus), cockroaches, butterflies, and crickets (Fig. 32). In addition, insects of the High Arctic seem to have a higher fat content, and thus a higher caloric content per gram, than insects from lower latitudes (Hoffmann, 1976). The chief body constituent of flies on Spitsbergen appears to be fat, and a figure of 5.41–5.73 kcal per g dry weight has been reported. On the other hand, the main reserve material of flies from temperate latitudes is protein and amounts to a total caloric content of 4.52 kcal/g. Even animals of the same species exhibit such differences: Phormia terrae-novae from Spitsbergen differs from laboratory individuals from Germany. It is too soon to decide whether such differences in body composition are simply modifications or whether genetical factors are at play. Studies on crickets and amphipods reared at different temperatures resulted in nothing comparable to the above dramatic differences in composition (Hoffmann, 1973; Teckelmann, 1974). As the pole is approached, remineralization by fungi and bacteria slows down due to the low and constant temperatures (Fig. 26). The differences between the Q_{10} of photosynthesis and that of the catabolic processes in poikilothermic organisms shown in this figure

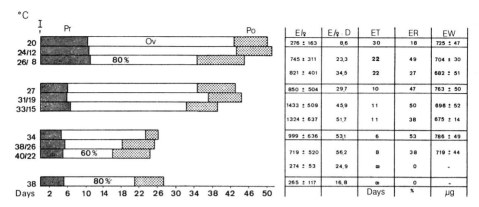

Fig. 32. Fertility of Gryllus bimaculatus under constant and under diurnally alternating temperature conditions. *I* imaginal metamorphosis; *Pr* period of preoviposition; *Ov* period of oviposition; *Po* postovipositional period; *E* number of eggs per female; *E/T* number of eggs per individual and day; *HT* time up to hatching of larvae; *HR* hatching rate; *HW* hatching weight. (After Hoffmann, 1974)

provide the basis for a wealth of inferences (see Remmert, 1978). Deviations from this scheme may occur in continental arctic regions with little rainfall and with high solar irradiation: animals inhabiting such regions can warm themselves up in the sunshine and attain a body temperature far above that of their surroundings. If, in addition, the local climate is particularly favourable, southern forms can penetrate astonishingly far to the north. The North Atlantic current is responsible for relatively high summer temperatures in northern Scandinavia, and continental areas within this region may even attain very high temperatures in summer despite their being far to the north of the Arctic Circle. This explains why lizards (Lacerta vivipera) and adders (Vipera berus) are found there. It is readily understandable that two viviparous reptiles have succeeded in surviving so far north since their offspring, too, can immediately benefit from the sun's warmth. In 1970 Kevan and Shorthouse made a thorough study of this phenomenon in butterflies of the High Arctic. Boloria chariclea, for example, can raise its body temperature about 11° C above that of the surroundings. On a dark substrate the temperature elevation may amount to even 15° C within the space of a minute (Fig. 33). At the same time the cyclic changes in body temperature essential for high egg-production are provided, so that the rate of reproduction may be greatly increased. The caterpillar stage, however, still presents a problem, and so far exact observations are available only on species from temperate regions. Its body temperature might possibly be raised by the uptake of heat radiation. Obviously such species can only exist in the continental regions of the Arctic (see p. 156).

Only very few poikilothermic animals in the Arctic are in a position to utilize the heat arising from a high rate of turnover (as in the rapid twitching of flight musculature) for their development (Figs. 34, 35). The bumble-bee (Bombus) has managed to bring off such a feat: it warms the hive by buzzing over it and thus distinctly accelerates the development of the larvae. This obviously requires a very substantial supply of nourishing, energy-rich food in the form of an abundance of

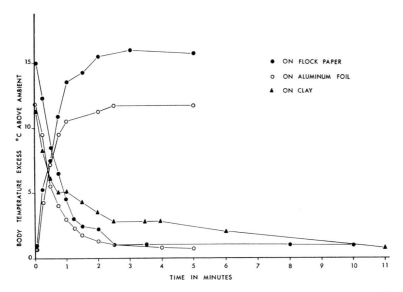

Fig. 33. Warming and cooling times of Boloria chariclea on different substrates: on black flock paper; on aluminium foil; on clay clod. (Kevan and Shorthouse, 1970)

nectar-filled flowers (Fig. 36). Again, this is only possible under continental arctic conditions with high summer irradiation (Figs. 37, 38).

The bumble-bee, like hibernatory mammals, is furthermore equipped with a mechanism serving exclusively to produce heat. In non-flying animals fructose-6-phosphate is converted into fructose-1,6-diphosphate with the consumption of ATP, and immediately broken down to fructose-6-phosphate and phosphate. The last process is accompanied by the production of heat. This is an example of a short cut in the metabolic cycle and is only seen in non-flying animals. In flying animals the fructose-1,6-diphosphate is converted into triose phosphate and this to pyruvate, which is the chief source of energy for flight (Newsholme et al., 1972; Hoffmann, 1978a). In order to utilize its own energy the animal has to be well insulated with regard to temperature, and this is very effectively achieved in the case of the bumble-bee.

The constant temperatures in the Arctic are also the explanation for the almost complete absence of frost during the growth season. Although snow may sometimes fall it melts rapidly and is never accompanied by frost. Insects of the High Arctic are sensitive to frost during the short arctic summer and quickly die in a freezer (Remmert and Wisniewski, 1970). Block and co-workers, on the other hand, were able to demonstrate a very large degree of frost-resistance in Collembola and mites of the Antarctic islands. The mites (Alaskozetes) appear to employ glycerine as an "anti-freeze" agent, although the frost resistance is only effective following evacuation of the intestinal canal. Glycerine accounts for 1% of the live weight of these mites.

The relatively constant low summer temperatures also have an effect on warm-blooded animals. This becomes particularly obvious in connection with the

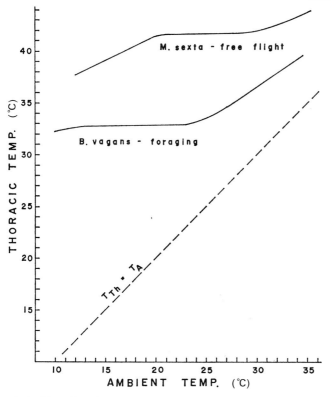

Fig. 34. Mean thoracic temperatures of two unrestrained insects as a function of ambient temperature during "normal" activity. The sphinx moth M. sexta regulates T_{Th} during free hovering flight with an approximate 5 °C range of temperatures by shunting excess heat from the thorax at $T_A > 20$ °C. At $T_A < 20$ °C and during flight at a lower energy expenditure than free flight (while supported on a flight mill), thoracic temperature varies nearly passively with T_A. In contrast, the T_{Th} of the bumblebee workers, B. vagans, during foraging (on-off flight) from Epilobium angustifolium, is regulated at the lower but not at the higher T_A. The bees produce heat at $T_A < 24$ °C while on the flowers, thus maintaining T_{Th} high enough for continued flight. At the higher T_A additional heat is no longer produced for the regulation of T_{Th} while the bee is perched on the flowers and T_{Th} then varies passively with T_A. The magnitude of the difference between T_{Th} and T_A then presumably varies with the percentage of flight, and possible variations of flight effort. (Heinrich, 1973)

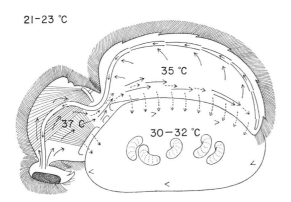

Fig. 35. Diagrammatic representation of the sagittal section of a queen bumble bee incubating her brood clump at room temperature. The directions of heat flow (---) from the thoracic muscles to the poikilothermic larvae (or eggs, pupae) which are being incubated, and the direction of blood flow (——) are indicated by *arrows*. (Heinrich, 1973)

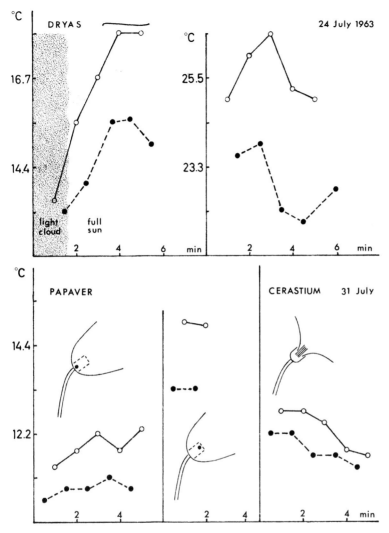

Fig. 36. Representative examples of temperatures recorded inside (○) and outside (●) heliotropic flowers (Cerastium, control) at L. Hazan, 1963. (Hocking, 1968)

breeding of birds, where a poikilothermic object has to be kept warm at a constant temperature over a long period of time by a warm-blooded one. A heavy burden rests on the warm-blooded partner in this process, and the lower the surrounding temperature the heavier it becomes. That certain adaptive strategies have been adopted can be safely assumed.

Under arctic conditions it is clear that species nesting in the open are at a disadvantage compared with those nesting in caves, and for this reason the latter can be expected to penetrate further polewards than members of the same group that nest in the open. This is exactly the situation and in fact the northernmost small

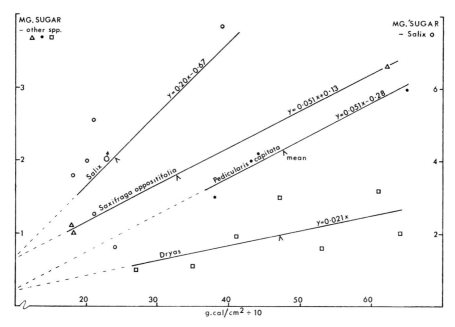

Fig. 37. The regression of nectar production in mg sugar per flower (catkin for Salix) on insolation for Salix arctica (○) male, Saxifraga oppositifolia (△), Pedicularis capitata (●), and Dryas integrifolia (□), L. Hazen, 1963. (Hocking, 1968)

passerines, the snow bunting (Plectrophenax) and the wheatear (Oenanthe), nest in caves. Considering the difficulties connected with feeding the young it would be expected that most species are nidifugous which, again, is the case. The number of bird species with nidifugous young very soon fending for themselves, is disproportionately high in the Arctic. Under certain circumstances this offers the possibility of rearing several overlapping broods of young. We may have hit upon a partial explanation of the much-discussed predominance of Laro-limicoles in the Arctic and of passerines in the tropics. Nevertheless, birds that nest in the open are confronted with very special problems, particularly the large species which have to brood longer. They are forced to lay their eggs on the cold ground before the growth season has begun. It is hard to imagine how the eggs can be efficiently incubated under such conditions. One way of alleviating the situation is to provide the eggs with very thorough insulation from the surroundings by means of an elaborate nest. This is what the passerines do, and their nests in the vicinity of the forest limits are invariably much more elaborate in construction and far better insulated than those of the same species in regions with milder temperatures. The thickness of the insulating layer in nests of the redwing (Turdus musicus) increases greatly from south to north. The same rule applies to the nests of the magpie (Pica pica). A classical example is provided by the eider duck (Somateria mollisima), whose eggs are provided with a thick pillow of high-quality down as protection against the cold ground. However, these are merely isolated examples: the overwhelming majority of arctic birds lay their eggs directly on the ground, e.g., guillemot, auk, fulmar

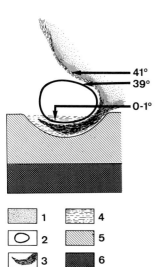

Fig. 38a. Temperatures in °C on the upper and lower sides of the egg of a Brunnich's guillemot brooding on melting ice. (Redrawn after Belopolski, 1957). *1* Body of the guillemot; *2* egg; *3* foot of the guillemot; *4* melt-water; *5* ice; *6* rock

(Fulmarus glacialis), tern (Sterna), skua (Stercorarius) and many limicoles. Eggs may even be laid directly on to frozen ground. Belopolski (1957) describes a particularly interesting situation in the nesting places of Brünnich's guillemot (Uria lomvia) on the Murmansk coast. He reports that the bird lays its eggs directly on to frozen ground in spring. Although during brooding the eggs may be taken on to the bird's feet they finally lie in ice water, which means a large temperature gradient (Fig. 38a). On the lower side of the egg the temperature is 0–1°C whereas on the upper surface it is about 39°C. When the bird settles on to the egg it pushes it with its beak until it is in contact with the bare skin of the brooding patch but otherwise completely surrounded by feathers. This usually happens in cold weather or if the birds have laid their eggs on the edge of snow-covered cliffs, which they often do in northern regions. The egg is frequently turned with the help of the beak to ensure reasonably uniform incubation. Despite this the bird is unable to compensate for the gross difference in temperature between upper and lower sides of the egg. According to Belopolski this large difference is even essential for the normal development of the embryo: in incubators in which temperature differences of this kind cannot be replicated, chicks only hatched from eggs that had been incubated for 28–30 days under natural conditions. If the eggs were put into the incubator earlier than this the embryos died (Belopolski, 1957). These results are sensational, and the experiments should be repeated as soon as possible, although so far this has apparently not been done. Even if the eggs do not require, but merely tolerate, such temperature gradients it would be a remarkably interesting adaption to life under arctic conditions. The breeding behaviour of the ivory gull (Pagophila eburnea) and the Antarctic snow petrel (Pagodroma nivea) requires further investigation. Both appear to nest on the frozen ground, and frequently make their nests far from the coast on nunataks. It is known that broods of Pagodroma nivea are lost due to freezing of the eggs (see Murphy, 1936).

The newly hatched young tolerate large fluctuations in temperature at first. The young of Brünnich's guillemot do not become endothermic until about the sixth

Fig. 38b. Many arctic birds build elaborate nests providing excellent insulation against cold. *1* Redwing (Abisko); *2* magpie (Tromsö); *3* Eider duck (Spitsbergen); *4* pink-footed goose (Spitsbergen)

day (Johnson and West, 1975). The young of the sandpiper (Calidris alpina) are obviously still partially ectothermic at first and are thus able to tolerate large fluctuations in body temperature. Many young birds are dark in colour, which probably helps them to exploit the warmth of the sun's rays better than their more southerly relatives [(skua (Stercorarius), long-tailed duck (Clangula), grouse (Lagopus), guillemots (Alcidae), sandpiper and relatives (Limicolae)]. In arctic mammals, too, endothermy develops during the first days of life. Although reindeer and caribou (Rangifer) calve in the spring, at a time of year when cold rain and snowstorms are frequent, the young animals are born without any truly protective hair covering and with no insulating fat layer. It is difficult to understand how they survive. At the age of 14–18 days lemmings (Lemmus) and voles (Microtus) are already able to maintain their body temperature for about an hour despite an environmental temperature of approximately $0°$ C.

Schmid (1972) has drawn attention to a phenomenon that had hitherto been overlooked: due to the large temperature differences between a warm-blooded body and the cold air of the Arctic the animal loses considerable quantities of water from its body surface by evaporation. A large animal loses relatively less water as a result of its relatively smaller surface area.

On the permanent snow and ice of mountainous regions of temperate and tropical latitudes, characteristic forms are encountered which constitute the so-called glacier flora and fauna. Specific algae often colonize regions of this kind, giving a blood-red colour to the snow (e.g., Chlamydomonas nivalis). The Fauna consists chiefly of Collembola and tardigrads. Drifting organic material, often pollen, darkens the glaciers and permanent snow to varying degrees, so that they are by no means always a pure white. This organic material, together with the algae, represents the food of the glaciers fauna. A similar flora and fauna is not known for the glaciers of the Arctic and Antarctic. Small quantities of glacier algae occur, but these are invariably found in water during their actual growth season, which is in the summer. There is no glacier fauna: one reason for this is probably the small amount of solar radiation, which is insufficient to warm up either plants or animals. In addition, the food at the disposal of animals is very limited since far less organic material drifts in than in the Himalayas or in the Alps. The glaciers and regions of pernament snow in the Arctic thus have no characteristic life of their own. Neither can the typical winter snow fauna of temperate latitudes be found in polar regions. This is not surprising in view of the fact that such a fauna consisting, for example, of Boreus and Chionea, undergoes its larval development during the warm, temperate although the imagine stage is a winter animal that can exist practically without any food. The arctic summer is too short for the larvae: Boreus occurs only up to the Hardangervidda and Chionea up to the Arctic Circle (Messaure, see p. 171 ff.).

In the permanent ice of the Antarctic Ocean, however, there is a regularly encountered fauna of unicellular algae (invariably diatoms). Although it is of regular occurrence it should probably not be regarded as typical. These forms can photosynthesize and even reproduce in the ice, which then becomes crumbly and finally releases the algae into the water (Buinitsky in Dunbar, 1977). Whether or not such phenomena can be observed in the northern polar regions is as yet unknown. In the Arctic genuine permanent marine ice cannot be expected at such low latitudes as in the Antarctic and the conditions required for photosynthesis, i.e., adequate light penetrating the ice, are less likely to be encountered at the higher latitudes.

Conditions in the arctic winter are very different from those already discussed in connection with its summer. Again, temperatures are uniform for relatively long periods of time and exhibit no particular diurnal rhythm. The lowest temperatures are by no means extreme and far lower values can be recorded in temperate continental regions, such as the interior of the continents of America or Asia, than on Spitsbergen. What was earlier presumed to be the cold pole of the earth is thus situated outside what we have defined as the Arctic. Survival in winter cold is obviously not a problem specific to the Arctic.

A covering of snow often greatly assists warm-blooded animals in the winter. Black grouse (Lyrurus tetrix) and species of Lagopus dig themselves into holes in the snow, from which they may suddenly fly out if they sense danger (Fig. 39b). Lemmings and other small mammals spend most of the winter under the snow, and the former are even able to reproduce under such conditions. Weasels (Mustela), too, largely disappear under the snow in winter although remaining active. Polar bear cubs are invariably born in snowy lairs, which are on the whole occupied from October until March. The one-year-old females and young males often hollow out and occupy similar quarters at this time of year although older males relatively seldom spend the winter in lairs of this kind. The large majority of these snowy lairs are situated on land, and relatively few have been found on the ice. They are particularly concentrated in specific areas such as east Greenland, east Spitsbergen, some groups of islands in the Soviet parts of the Arctic Ocean, and on certain islands of the Canadian Archipelago. Although still very low, the temperature inside the lairs is higher than on the surface of the snow (temperatures of $-9.9°$ and $-17.8°$ C have been recorded, for example). This means that there is a difference of $21°$ or $7.8°$ C, respectively, as compared with surface temperatures. Unfortunately

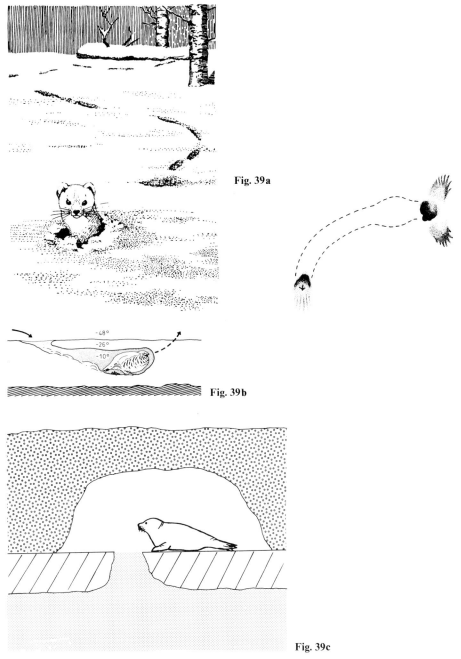

Fig. 39a-d. Events rarely observed in the true Arctic due to insufficient snowfall (glaciers would form if there were this much snow): small animals "dive" in the snow or rest in snowy lairs. **a** A weasel and its tracks in the snow. **b** Diagrammatic representation of the hide-out of the hazel hen. The entrance and the hole through which the bird flew out of the snow can both be seen. In winters with little snow the birds freeze despite adequate food supplies. Observations from Siberia, south of the Arctic Circle. (Bergmann et al., 1978). **c** Snowy lair of the ringed seal. **d** (p. 44) Polar bear den. (Compiled after various authors)

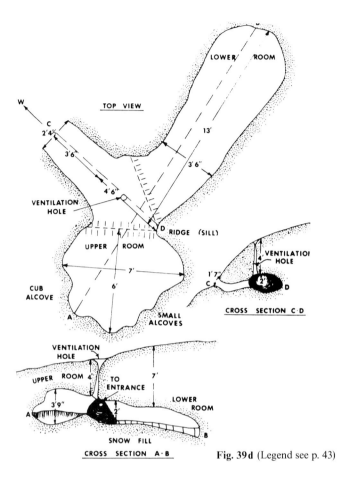

Fig. 39d (Legend see p. 43)

it is not known whether the normal temperature of the snow at the same depth might not be the same as in the lairs, which would seem to be quite feasible. It appears that alterations and extensions are made throughout the winter (Harington, 1968), and at all times there is good ventilation shaft (Fig. 39d). Ringed seals often spend the winter in similar quarters. They make a breathing hole in the ice, keeping it open even when snow falls, so that eventually a snowy den is formed on the ice above it. The animals now have a relatively warm shelter, due to its direct connection with the open water and the excellent insulating properties of snow. The polar bears have developed a special hunting strategy for detecting these seal lairs in winter. They take a jump and come down heavily on the snow with their forelegs. If they land on a seal den the snow collapses, the seal has no time to escape and is an easy prey. A covering of snow, however, presents problems for some animals, as for example birds that cannot find the grit necessary to break down food in the gizzard. This is often a greater hazard to their survival than scarcity of food or cold (Meinertzhagen).

Notably in the continental regions of the Arctic, "snow hollows" play a special role for both flora and fauna. Snow is blown into these small hollows by the wind,

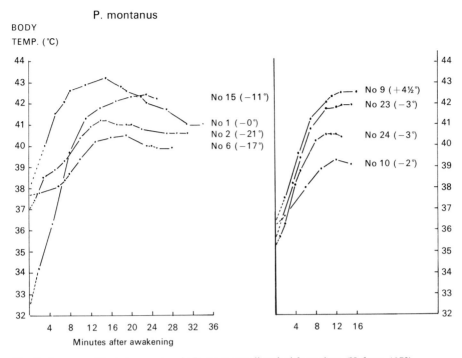

Fig. 40. Increase of body temperature in P. montanus disturbed from sleep. (Haftorn, 1972)

so that even in regions with little snowfall they always provide a protective covering for plant species sensitive to frost and the accompanying desiccation. Examples of typical snow hollow species on Spitsbergen are the dwarf birch Betula nana, the berry-bearing Rubus chamaemorus and Cassiope tetragona, a species that suffers severe frost damage in the absence of snow cover. In springtime when the sun's rays become stronger again light penetrates the thin layer of snow in these snow hollows to the dark ground beneath and causes thawing to begin from below. A kind of hothouse develops under the snow, with relatively high temperatures and sufficient light for photosynthesis to achieve a positive balance. Under these conditions animals can also become fully active. Thus the snow hollows represent refugia, particularly for warmth-loving animals in exposed areas (on Spitsbergen, for example, for the beetle Rhynchaenus flagellum); and their vegetation (with Salix species) provides warm-blooded herbivores with an important source of food.

This brings us to the question of what the typical arctic winter involves for living organisms. During this season only warm-blooded organisms are capable of activity on land, and they can be distinguished according to the times of the day at which they are at rest or fully active.

Apparently a number of passerine birds lower their body temperature whilst resting during the winter nights in order to save energy. This was demonstrated by Haftorn (1962) in certain tits (Parus) of the marsh-tit group that pass the winter in Lapland (Fig. 40). They spend the long nights in caves, and their body temperature drops to about 32° C, while during activity it is 41°–42° C. During metabolic rest the

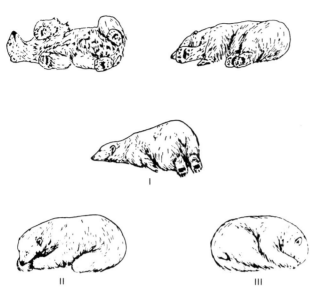

Fig. 41. Polar bears' postures at mean windchills 830 W/m² (*I*), 1410W/m² (*II*) and 1910 W/m² (*III*). (Øritsland, 1970)

temperature is probably held at about 39° C. Unfortunately, scarcely any other reports from genuine arctic regions are available. At best, the data refer to places in the immediate vicinity of the Arctic Circle, and were not obtained during the darkest time of year. In the regions where such investigations were carried out the animals can as a rule take up sufficient food since the days are relatively long, and it is no longer so important for them to lower their temperature at night. This led West (1974) to put forward the hypothesis that principally the species that find adequate and energy-rich supplies of food remain over winter. This applies to guillemots (Alcidae) for example, which are relatively poorly insulated from the environment (see p. 126 for conditions under water). It applies equally to Carduelis flammea, the redpoll, which shows no signs of torpor and has large quantities of energy-rich food at its disposal.

Faced with extreme cold, resting animals adopt an increasingly spherical position, thus achieving a relative reduction of surface area, which lowers heat losses (Fig. 41). Emperor penguins (Aptenodytes forsteri) that breed in the antarctic winter huddle together in a dense group during snowstorms, with the same effect (Fig. 42). Sleeping communities of passerines during the winter probably serve the same purpose, and similar communities of small mammals have also been reported.

However, temperature values alone convey very little information and should invariably be considered in conjunction with air movements. In most arctic regions there is little protection from wind due to the absence of trees. Places that would provide shelter from the weather, behind rocks and mountains, are usually covered by deep snow, making it difficult for herbivores to find food. Air movements consume heat, especially if the air contains snow or ice crystals (Table 3). The best performance seems to be put up by the emperor penguin of the Antarctic, which

Table 3. The cooling effect of the wind. The temperatures within the thicker line are hazardous to man: flesh freezes within 1 minute. Temperatures to the right are extremely hazardous; flesh freezes within 30 s. At a temperature of $-29\,°C$ and a wind speed of 40 km/h, humans are affected as they would be by $-60\,°C$ in still air. (Weiss, 1975)

Wind velocity in km/h	Temperature in $-°C$														
	9	12	15	18	21	24	26	29	31	34	37	40	42	45	47	51
8	12	15	18	21	24	26	29	31	34	37	40	42	45	47	51	54
16	18	24	26	29	31	37	40	42	45	51	54	56	60	62	68	71
24	24	29	31	34	40	42	45	51	54	56	62	65	68	73	76	79
32	24	31	34	37	42	45	51	54	60	62	65	71	73	79	82	84
40	29	34	37	42	45	51	54	60	62	68	71	76	79	84	87	93
48	31	34	40	45	47	54	56	62	65	71	73	79	82	87	90	96
56	34	37	40	45	51	54	60	62	68	73	76	82	84	90	93	98
64	34	37	42	47	51	56	60	65	71	73	73	82	87	90	96	101

even breeds in the antarctic winter (see p. 94). Conditions in the Arctic are not as extreme, but even here the high wind velocities can cost energy, as Fig. 43 shows for the snowy owl.

In the arctic winter marine birds and mammals have to cope with the special problem that their fur or feathers and extremities freeze as soon as the animals emerge from the water. For this reason certain arctic birds such as the ivory gull (Pagophila eburnea) and the roseate gull (Rhodostethia rosea) avoid getting their feet and feathers wet in winter and try to catch their food on the water surface.

This possibility is not open to birds and mammals that have to dive for their food. An insulating layer of air among the hairs or feathers would make diving much more difficult, and so the water penetrates right down to the animal's skin. The "insulating" layer of hair or feathers thus becomes ineffective (Fig. 44). The only help in such cases is the presence of a thick layer of subcutaneous fat (Fig. 45), and/or a greatly elevated rate of metabolism, accompanied of course by a considerable increase in food consumption. On land, however, the insulation provided by fur or feathers is usually sufficient to render substantial elevation of metabolism unnecessary (Fig. 46). Just as in water, however, the protruding parts of the body like feet, ears or snout are a problem. These extremities can be considered to be largely ectothermic. The composition of their membrane lipids differs from those in the main part of the body (more unsaturated fatty acid, which means higher fluidity): the essential metabolic enzymes function even at very low temperatures (Irving, 1972; Hochachka and Somero, 1973).

Simple loss of heat due to cold and air movements is in general well tolerated by animals in polar regions. Those functions are endangered, however, in which highly sensitive thin-skinned parts of the body come into immediate contact with the icy

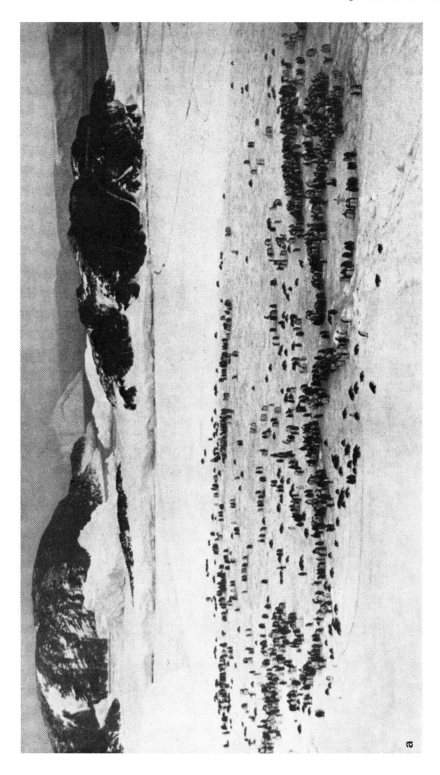

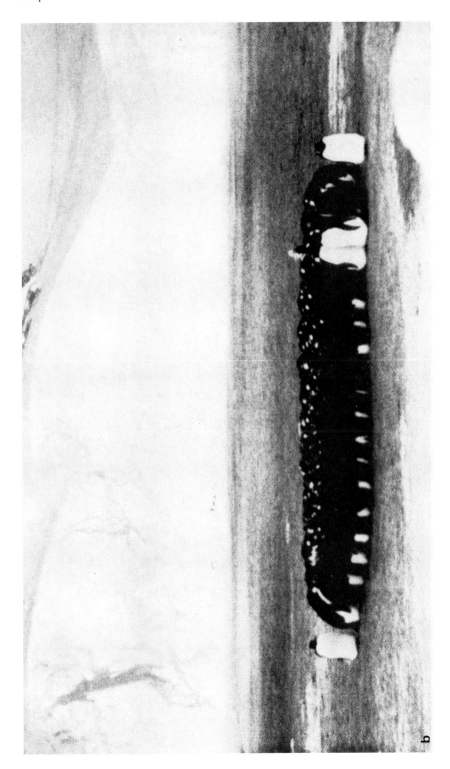

Fig. 42. Distribution of emperor penguins in a colony in good weather **a** and during a snowstorm **b**. Diagrammatic, after photographs by Rivolier (1957)

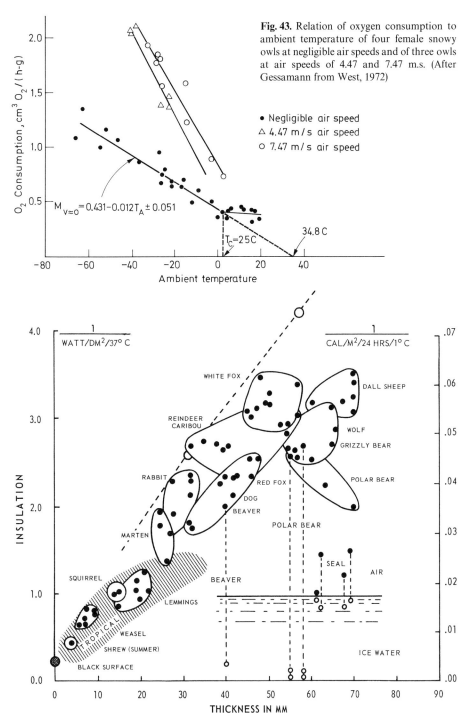

Fig. 43. Relation of oxygen consumption to ambient temperature of four female snowy owls at negligible air speeds and of three owls at air speeds of 4.47 and 7.47 m.s. (After Gessamann from West, 1972)

Fig. 44. Insulation in relation to winter fur thickness in a series of arctic mammals. The insulation in tropical mammals is indicated by the *shaded area*. In the aquatic mammals (seal, beaver, polar bear) the measurements in 0 °C air are connected by *vertical broken lines* with the same measurements taken in ice water. The *two upper points* of the lemmings are from Dicrostonyx, the others from lemmus. (Scholander et al., 1950)

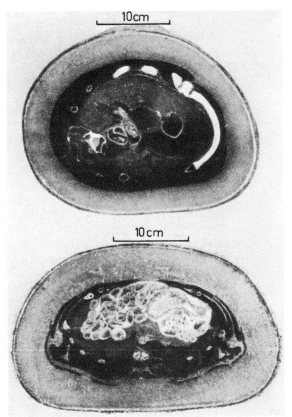

Fig. 45. Cross sections of two frozen seals (Phoca hispida) shot in March, 1948, showing the thick layer of blubber. (Scholander et al., 1950)

surroundings (Fig. 47). Such dangers are involved, for example, in urination or defaecation (the excreta freeze to ice immediately upon leaving the body); in feeding (thin-skinned parts of the mouth cavity are in immediate contact with the outside air, and in the case of herbivores with very cold food); and most of all in respiration. The eyes with their sensitive cornea are also continually exposed to danger. The two first-mentioned functions can be suppressed for a few days during exceptionally bad weather, but respiration goes on continuously and the highly delicate lungs are constantly being filled with air from the surroundings. If the snout is no longer adequately warmed drastic damage occurs and death ensues immediately. So far detailed experimental studies of this phenomenon have not been carried out and it is a factor that has until recently been neglected. Boreal owls (Aegiolus funereus) are even capable of thawing frozen mice by warming them in their plumage near the skin (Scherzinger, 1979). Furthermore, the energetic value of the food taken during severe cold is decreased by a factor that depends on the digestibility of the food and the energetic costs of heating the cold food. To compensate, food intake must be appropriately increased.

Poikilothermic organisms exhibit no active signs of life below freezing point (land and freshwater: $0°C$, sea water: $-1.7°C$. Only a very few species tolerate

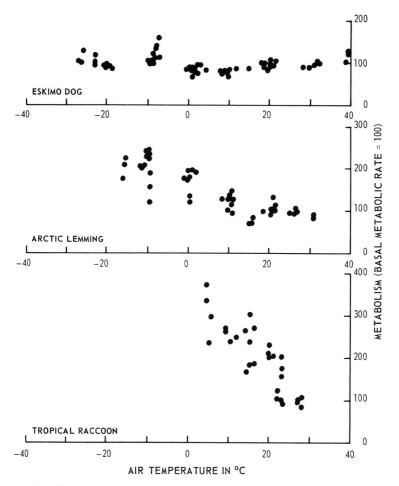

Fig. 46. Effect of environmental temperature on the metabolic rate in Eskimo dog, arctic lemming, and tropical raccoon expressed in terms of basal metabolic rate = 100. The steepness of the curves describes the relative temperature sensitivity of the animal but does not correlate with the weight of the animals or with their body insulation. (Scholander et al., 1950)

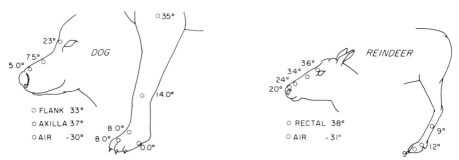

Fig. 47. Topographic distribution of superficial temperatures in a dog and in a reindeer. (Irving and Krog, 1955)

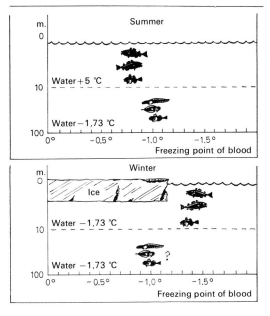

Fig. 48. The freezing point of the blood plasma of shallow water fish (Gadus ogac, Myxocephalus scorpius, above broken line) and of the deep-water species (Lycodes, Liparis, Gymnacanthus) in summer and winter. The position of the fish on the abscissa indicates the freezing point of the blood plasma (After Scholander et al., 1950 from Theede, 1967)

intracellular ice formation and therefore freezing of the body fluids has to be prevented at all costs. This is achieved by the addition to the body fluids of anti-freeze agents which can act in one of two ways. One possibility is that both freezing point and melting point are depressed, a mechanism usually only effective at relatively mild temperatures (Fig. 48). It requires the solution of a large number of small particles in the body fluids, thus raising the osmotic pressure and considerably changing the internal milieu of the organism. Consequently, and especially at very low temperatures, a different strategy is adopted. Organic substances, of which glycerine has become best known, are added to the internal milieu to prevent ice formation. The anti-freeze agents of marine fishes are without exception glycoproteins of widely differing molecular weights (3000–32,000 daltons). The most important among them are made up of the two amino acids threonine and alanine in the ratio of 1:2, plus the sugars N-acetyl-galactosamine and galactose. It is therefore a complicated mixture of chemically similar substance that is responsible for lowering the freezing point. However, it is possible that the situation in the Arctic is different from that in the Antarctic. No macromolecular anti-freezing substance could be demonstrated in the serum of the arctic-boreal fish Myxocephalus scorpius, although two peptides with a protective action against frost could be isolated from its skin. These might account for the relatively high frost resistance of the species by rendering the skin an effective barrier to the growth of ice crystals. Contact with ice does not elicit freezing of the super-cooled body fluids of this species. The protective peptides isolated from M.scorpius consist solely of 11 amino acids. Both peptides contain a strikingly high proportion of alanine, which is characteristic of all anti-freeze proteins so far known from fish (summary: Schneppenheim, 1978). In such cases freezing point and melting point are very different. Armed with a similar mechanisms arctic beetles can safely tolerate temperatures as low as $-80°$ C.

Marine teleosts are faced with a particular problem. As descendants of ancestors that lived in the freshwater-terrestrial sphere they are secondary marine forms (Remmert, 1968, 1978) and the osmotic pressure of their body fluids is lower than that of the surrounding ocean. This means that their body fluids freeze before the water in which they live. If ice crystals enter their gills, as often happens during stormy weather, the animals may become frozen stiff within a very short time. Arctic fish and fish from other equally cold regions like the Baltic thus either have protective substances in their body fluids or migrate to deeper layers of water. In primary marine animals, however (in species that at no time during phylogenesis had freshwater or terrestrial ancestors), the osmotic pressure of blood and tissues is slightly higher than that of the medium in which they swim. They freeze at lower temperatures than their medium and are still able to escape before the sea water freezes. The significance of low temperatures is therefore very different for teleosts, on the one hand, and for starfish, sea-urchins, marine molluscs and marine crustaceans on the other.

The plants and animals of the marine tidal regions in the Arctic are also faced with special problems. Twice a day they have to tolerate submersion in water whose temperature is about $-1.7°$ C and subsequent exposure to air of possibly very low temperatures. This, in addition to the effects of drifting ice, probably explains the very sparse macrofauna of the marine littoral of the High Arctic. Its microfauna has a specific composition, with a very low proportion of small crustaceans.

One more event to be borne in mind, although its biological significance is not yet clear, is that in marine shallow-water regions (up to about 15 m depth) ice may form on the bottom due to the differences in density between freshwater and sea water.

We shall now turn our attention to problems arising in connection with permafrost. It is known that the temperature in deep caves is very constant and equals the annual mean temperature of the locality concerned. The same applies to the temperature of the soil. Consequently, wherever the annual mean temperature is below freezing point the ground ought to be permanently frozen, i.e., it should lie in the permafrost region. Only at depths approaching 300 m, which are of no biological interest, is the ground in these regions assumed to be thawed by heat flow from the earth's interior. Although this principle does apply within broad limits there are many extensive areas where it apparently does not. Deviations can be explained roughly as being due to long-term fluctuations in climate and to the fact that the temperature in the ground adjusts very slowly to changed climatic conditions. The areas of permafrost still existing in the Swiss Alps are undoubtedly remnants of the glacial epoch: an unusually sheltered situation prevented the normal process of thawing. Spruce grows very slowly in such localities and remains stunted.

Permafrost today covers almost the entire region which we have designated as the Arctic. Northern Scandinavia is the only part of it with no more than a few isolated permafrost areas, which are concentrated on moors. In central Asia permafrost stretches as far south as central European latitudes (Fig. 2).

Very little is known concerning the biological significance of permafrost. A few possible consequences will be discussed in the following—mostly based, however, on theoretical considerations and with little support from field observations.

Only one mammal is known to hibernate in the arctic regions of North America and this is the ground squirrel (Spermophilus undulatus), so that a far smaller percentage of mammals hibernates in the Arctic than in the temperate continental regions (Fig. 49). Obviously this one species occupies a special position among hibernators. In temperate latitudes hibernating animals seek a relatively warm hideout in the ground where frost cannot reach them, and then proceed to insulate it thoroughly from the surroundings. The animal's body temperature remains at about 8 °C throughout the winter. This is impossible in the perenially frozen ground of the Arctic. If the animals were to retreat to a cave they would be confronted with temperatures below 0° C, and it would be much more difficult for them to maintain a body temperature of 8° C than under the conditions described for temperate latitudes. Even if they were to resort to intensive insulation of the cave with plant matter the situation would not be essentially improved. Warming up of the walls of the cave would lead to water and ice formation, which could be dangerous. This is a point that has not so far been satisfactorily resolved. We do not know how the winter quarters of the arctic ground squirrel compare with those of the same species living further south, nor do we have any information about the temperature conditions in such a lair. There is still no evidence as to whether the arctic ground squirrel begins the winter with much larger reserves than its southerly relatives, and whether it perhaps chooses regions with coarse, porous material in which to make its winter habitation.

The same problem applies to poikilothermic animals overwintering on land in a resting state. In southern latitudes frost tolerance is not obligatory because the animal is able to avoid the cold either by digging itself deeply into the ground or by seeking shelter in some well-protected spot. In permafrost regions this is clearly not feasible, so the animals are obliged to be frost-resistant in order to survive. It ought therefore to be possible to detect a sharp faunistic boundary, for poikilotherms at any rate, between permafrost and unfrozen regions. (The overwhelming majority of poikilothermic animals of temperate latitudes cannot tolerate frost.) Data on this point are not yet available.

Even less information is available about limnic ecosystems during the arctic winter. Most running water disappears from permafrost regions during the winter because the springs can no longer be fed, and how the organisms of such localities pass the winter remains an enigma. Vast numbers of small bodies of water freeze to solid ice in winter but here, too, the fate of the living organisms is unknown. And lastly, there is the question of what happens in the larger bodies of water that do not freeze completely, and whether (and if so, how) ice also forms from the bottom upwards. Most of the continuously flowing water in the Arctic is carried by rivers that arise in the south and are merely traversing the permafrost region. Their animal life has not yet been adequately studied and still less is known about the conditions prevailing in winter. Spring is often associated with problems in these rivers. Whilst the snow is melting fast in the south a thick layer of ice may still cover the northern reaches. The vast quantities of water can only be coped with by means of an increased velocity of flow, although this cannot be raised indefinitely. Under these

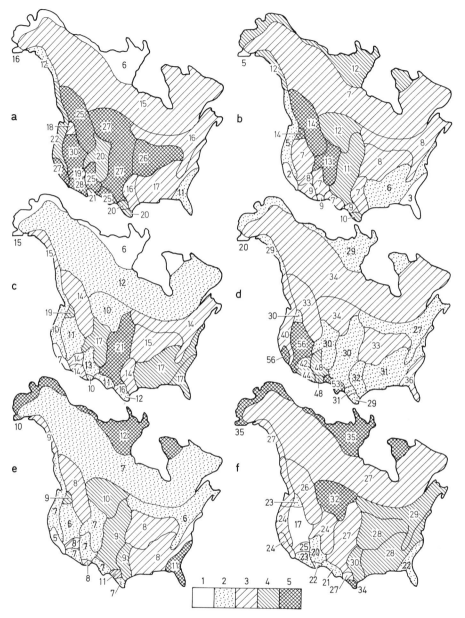

Fig. 49. a Percent of mammals undergoing dormancy; **b** percent of fauna made up of large herbivores; **c** of medium herbivores; **d** of small herbivores; **e** of large carnivores; **f** of medium carnivores. (After Hagemeier and Stulz, 1963)

conditions the ice is often subject to such high pressure from below that, even in places where it is almost a metre thick, the water may burst through. Thus, in the middle of the arctic winter a river may quite suddenly have open water. No one has yet studied the effect of such pressures on the swim bladder of fish. Another place where flowing water is always found is at the exits from the large arctic lakes. In some places they may even remain free of ice at a latitude of 80°N (see p. 213). Thus in summer there are several almost indistinguishable types of running water in the Arctic: the relatively warm rivers coming from southern regions, the relatively cool rivers that drain arctic lakes, and the rivers that carry no water in winter. It is logical to assume that very large differences exist between the populations inhabiting them, although no observations have been reported so far. The extensive review of Zhadin and Gerd (1963) can be regarded as a basis for studies that still have to be carried out. Neither in the large section on nordic and arctic rivers, nor in that dealing with arctic lakes is the problem of permafrost given any consideration, and no mention is made of the role of marine elements in certain rivers and lakes of the Arctic.

Phenomena of this kind are of special significance for the surrounding land. When the ice breaks the warming effect of the water on the immediate surroundings is considerable. As an example, the air temperature at 15^{10} h on 20 June 1973 in the central, northern Mackenzie Delta was 19.5°C and the water temperature 16°C. Only 5 km to the north-east of the delta the air temperature (20 min later) was 7°C, the lake situated at this point still had a complete covering of ice and the temperature of the small water courses in the vicinity was between zero and 0.5°C. The advancement of spring in the neighbourhood of the river is clearly reflected in the fact that the willows and other bushes turn green much earlier, and their buds open 2–3 weeks earlier than elsewhere in the surroundings (Gill, 1974).

Special problems are connected with perpetually frozen lakes. It would be reasonable to expect that they are frozen solid, and that they contain no living forms, being cut off from the surrounding by the permafrost ground below and the layer of ice above. Surprisingly, studies carried out in the Antarctic revealed that lakes of this kind can act as traps for solar energy. Sunrays hitting the surface at a low angle are refracted into the depths of the lake and raise the temperature of the water to about 8°C. Only a few decades ago a large antarctic glacier receded slightly, revealing the presence of a lake. Although this lake had never been without a covering of more than 1 m of ice, in which there were neither cracks nor splits, the water was found to contain blue and green algae as well as bdelloid rotifers. Nevertheless, production is extremely small. It is possible that lakes of this kind are non-existent in the northern hemisphere, but the observation shows that it is wise to be prepared for the unexpected, and that production is indeed possible beneath perennial ice.

Low water temperatures are not unique to the Arctic. They are encountered in an extraordinary variety of regions and in fact the winters on low land at 40°N can be colder than in northern Scandinavia at a latitude of 70°N. The mechanism of cold tolerance has not been dwelt upon at length since it is not a typically arctic attribute. The centre of distribution of hibernatory mammals (Fig. 49) lies in continental regions with warm summers and cold winters. Resistance to cold is not a privilege confined to arctic animals, as evidenced by the camel (Camelus bactrinanus). In Austria, Hungary, and North America the praying mantis (Mantis) tolerates winter temperatures far below those of the west coast of Norway or the southwest coast of Iceland, where its absence is due to the low summer temperatures. The coexistence of (arctic) snowshore hares and (arctic) caribous with (tropical) humming-birds in Canada and in the northern United

States points in the same direction. This is a fact that ought to be borne in mind when speculating about the reasons for the extinction of pleistocene mammals like the cave bear, woolly rhinoceros and mammoth. There is much to support the theory that these were not arctic forms at all, but were comparable to our present-day camel in requiring the high summer temperatures and long growth season that prevailed during the glacial period in Europe. Since polar displacement cannot have been responsible for the Ice Ages, there must have been intense solar radiation in these areas, a point that is given far too little attention in any discussion on the disappearance of pleistocene species.

In summing up, the situation can be described as follows: due to the constant low temperatures on land and in water, breakdown processes in the Arctic are slow (photosynthesis is less affected since it is partly a photochemical process). This means that all poikilothermic organisms whose metabolism is based on breakdown processes (bacteria, fungi, poikilothermic animals), grow slowly. Collembola, for example, take several years to develop in the High Arctic, butterflies up to 10 years; antarctic amphipods take 5 years to attain a length of 5 cm. Warm-blooded animals are at an advantage because primary production is high and food thus plentiful, which enables them to maintain the large difference between internal and external temperatures. Consequently, warm-blooded organisms play a much larger role in arctic (and antarctic) ecosystems, terrestrial and marine, than in any other environment of the earth.

3. Ecological Factors Other Than Temperature and Light

Light and temperature occupy a pre-eminent position among the ecological factors of the Arctic, every event being subordinated to them. The degree of their predominance is evidenced by the fact that hardly any further distinction between arctic animals due to other ecological requirements has become manifest. This explains why different animal communities have scarcely ever been established.

Obviously the barrier represented by the salt content of the sea has the same significance in the Arctic as elsewhere and it need scarcely be emphasized that the boundaries between freshwater and sea and between land and sea are still apparent. The very violent influx of freshwater into arctic seas caused by melting snow can lead to a sudden and marked reduction in salt content, particularly in coastal regions. The implications of such sudden changes have been relatively little studied. Skreslet (1973) assumes that they function as a zeitgeber controlling the expulsion of sexual material in the mollusc Chlamys islandica, possibly in conjunction with the equally sudden temperature changes.

The boundaries between freshwater and land are still recognizable in the Arctic, but any finer subdivision on the basis of factors other than light and temperature, or factors immediately dependent upon them, has not been seriously attempted and seems to be a difficult undertaking. The problems begin in the soil itself, which has a high content of humic acid as a result of the slow remineralization of plant matter. Thus, whatever the geological basis, there is a tendency to raw humus formation.

For this reason, the very large differences in density of mountain hare (Lepus) and grouse (Lagopus) in Scotland, depending upon whether the soil is acid or alkaline, cannot be expected in the Arctic. One exception to this is Iceland, with its favourable climate and recent volcanoes, where the rich alkaline soils support a correspondingly rich flora and fauna (see e.g., p. 203). But in fact Iceland is to the south of the Arctic Circle and thus beyond the limits set by our definition.

Soil humidity plays a considerable role and is to some extent correlated with temperature. Oceanic regions with frequent rainfall and little sunshine have lower summer temperatures than continental regions. However, the fact that such oceanic regions are apparently hostile to a number of arctic homoithermic species is certainly not due to the temperatures. A well-known illustration is provided by the muskox (Ovibos moschatus), whose distribution is limited to continental arctic regions. Every attempt to introduce it into oceanic regions of the Arctic (Iceland, north Norway) met with failure. To a lesser extent the same seems to apply to the reindeer (Rangifer): of all the colonies set up in Iceland only that in the interior (i.e., the most continental part of Iceland) is still in existence today, whereas those introduced into the oceanic regions have disappeared. It is difficult to offer an explanation at the moment for the deleterious effect of an oceanic climate on warm-blooded terrestrial animals. Their marine counterparts (penguins, guillemots, seals) are excellently adapted to such conditions. One argument is that severe hoarfrost forming on the animal's fur may lead to the loss of its insulating properties. Certain animal species such as the glutton (Gulo gulo) are said to be equipped with hairs on which ice crystals cannot form. But all of this speculation rather than the result of concrete observations.

Even in continental regions with strong solar radiation there are wet areas bordering running water. Their vegetation consists mainly of grasses and sedges, but scarcely any dicotyledonous flowering plants. The soil is inhabited chiefly by the larvae of terrestrial chironomids, but apart from this its fauna is very impoverished, as shown by catches in formalin traps. The dry areas are much richer in species: terrestrial chironomids play a relatively minor role here, whereas mycetophilids and sciarids are of importance. There are also large numbers of spiders in addition to the other members of a typical terrestrial arctic fauna (cf. e.g., p. 157 and Figs. 84, 119). In strongly eutrophic areas below bird rocks a large number of trichocerids can be found. Parasitic Hymenoptera occur everywhere in very large numbers. With these few observations we appear to have exhausted the possibilities of a further subdivision of habitats in the High Arctic on the basis of factors independent of climate.

The few herbivorous poikilothermic organisms of arctic regions (with the exception of northern Scandinavia) are not so strictly specialized with respect to their food plants as is generally the case in temperate or even tropical latitudes. The width of the niches of these species increases. Whereas in the tropics about 85% of the Papilionidae are specialized on certain food plants, only 50% of the caterpillars at 65°N exhibit this kind of specialization (Scriber, 1973, quoted from Kauri in Wielgolaski, 1975). Almost all species in the Arctic can be found on any of a large number of, or at least on several different food plants, their distribution mainly depending upon the climatic conditions (mostly in regions with relatively strong and regular sunshine). The terrestrial warm-blooded animals, on the other hand,

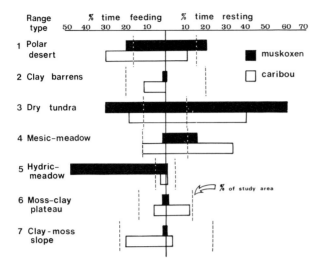

Fig. 50. The summer use of vegetation types by muskoxen and caribou at Bailey Point, Melville Island

appear to show definite specialization with respect to food. Reindeer (Rangifer), for example, are the typical form of the dry tundra, whereas the muskox (Ovibos) prefers to graze in sedgy swamp meadows (Fig. 50). Muskoxen thus depend upon the wet areas of the continental tundra regions for their well-being (Barker and Ross, 1976). On Spitsbergen and on the Dovre-Fjell, where the species has done well since its introduction (but see page 200), the above observation from its native regions could not be confirmed. Instead, the muskoxen prefer the steep mountainous slopes, whilst the reindeer keep to the flat dry tundra. Apparently the steep slopes are still slightly more continental than the wet areas of the mountains and the polar cattle find sufficient wet fodder to exist.

Where several species of Calidris occur together they, too, very possibly have different food preferences. For example, Nettleship (1974) demonstrated on Ellesmere Island that the knot (Calidris canutus) prefers the dry Dryas tundra for breeding and that its young chiefly feed on adult chironomids. Plant matter, on the other hand, makes up an important part of the food of the adult birds, and from the time they return from winter migration until mid-June it forms the basis of their diet. Even when the young birds are still being led, and almost exclusively catch adult chironomids, plant matter never accounts for less than 50% of the diet of the adult birds. The perigynia and achenes of Cyperaceae, the stalks of Equisetum, the buds of Polygonum and the sporophytes of mosses are consumed. It can safely be assumed that other Calidris species have completely different nutritional habits. The decrease in number of species from south to north is paralleled by other changes. The songs of the remaining species are less clearly definable (this is especially obvious in the case of the snow bunting on Spitsbergen, where it is the only song-bird and its song is of astonishing variability).

The variety of ecological situations under which the different arctic guillemots live have been the subject of very thorough studies (particularly by Belopolski, 1957; Kartaschew, 1960). The black guillemot (Cepphus), which lays two eggs,

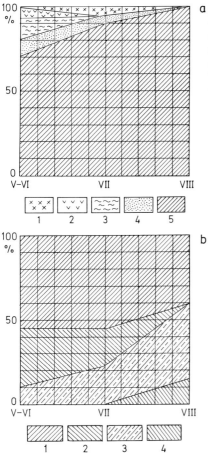

Fig. 51a and **b.** Seasonal changes in the proportion of various food species in the food of the puffin on the east Murmansk coast. (After Belopolski, 1957). **a** Proportions of all food species: *1* sponge; *2* insects; *3* polychaetes; *4* crustaceans; *5* fish. **b** Ratio of the various fish species in the food: *1* sand eel; *2* caplin; *3* herrings; *4* young cod

fishes in the immediate vicinity of the shore, whereas the razorbill (Alca torda), guillemot (Uria) and puffin (Fratercula) each lay only one egg and fly far out to sea to catch their food. The razorbills and guillemots accompany their solitary offspring to the water but the puffin, which spends its time much further out, leaves its young to find their own way down to the sea and swim away. The little auk (Plautus) generally fishes near the shore or near the edge of the ice. It is not surprising therefore that the food spectra of these species vary considerably, as shown in Figs. 51–54.

The sharp ecological distinction between the species is equally well maintained in their breeding colonies. The black guillemot breeds singly or in very small colonies, making its nest below small rocks. Puffins chiefly nest in hollows that they make themselves, and the little auk nests in gravel formed by the mechanical weathering of the rocks in the Arctic. Colonies of razorbills and two guillemot species (Uria aalge and Uria lomvia) are found on the steep coasts, the razorbills choosing broader strips of cliff than the guillemots. No ecological difference between the breeding places of the two guillemot species seems to be detectable. The

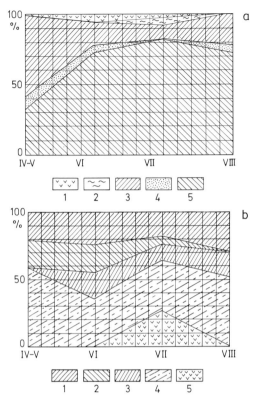

Fig. 52. Seasonal changes in the proportion of various food groups in the food of the black guillemot on the east Murmansk coast. (After Belopolski, 1957). **a** Proportions of all food species: *1* insects; *2* polychaetes; *3* crustaceans; *4* molluscs; *5* fish. **b** Ratio of the different fish species in the food: *1* sand eel; *2* caplin; *3* herring; *4* young cod; *5* other fish

two are found side by side in many colonies on the north coast of Norway and on Iceland, although it is mostly a case of smaller colonies of one species scattered within the main breeding territory of the other. Nowhere has competition between the two species been observed at feeding sites and it is unanimously agreed that it does not exist.

The food requirements of antarctic seals appear to differ very little from one species to the next. An exception is provided by the leopard seal. which lives on fish, birds and mammals, whereas the rest live mainly on krill. Whether the differences between arctic seals are equally small is not yet known. They seem to be mainly fish consumers, only the walrus showing a preference for benthic organisms which it fishes from the ocean bed. In some respects the walrus appears to fulfil the same ecological function as the sea leopard in the Antarctic. Given the opportunity it will kill and devour fish and seals. Even the polar bear apparently prefers to avoid a confrontation with a walrus and is usually conspicuously absent from regions with a high walrus population, or is at least rarely encountered. However, the behaviour of polar bears differs extraordinarily from one population to another: the Hudson Bay population mainly roams about the tundra during summer, consuming plant matter just like land bears. The polar bears of Spitsbergen and Franz-Josef's land, on the other hand, are purely carnivorous, living the whole year round on seals and thus sharing the role of the antarctic sea leopard with the walrus. It is probably too

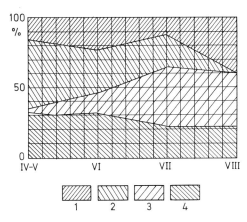

Fig. 53. Seasonal changes in the proportion of different fish species caught by the guillemot off the east Murmansk coast. (After Belopolski, 1957). *1* sand eel; *2* caplin; *3* herrings; *4* young cod

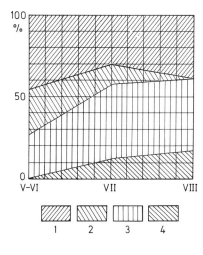

Fig. 54. Seasonal changes in the proportion of different species caught by the razorbill off the east Murmansk coast. (After Belopolski, 1954). *1* sand eel; *2* caplin; *3* herring; *4* young cod

late to determine the extent to which interaction between individual species, comparable to that suggested between walrus and polar bear, occurs or has occurred in the past. Reports that the walrus can attack the killer whale (Orcinus orca) and drive it from its revier suggest the possibility of highly interesting interactions.

Due to the absence of warm-blooded terrestrial predators in the Antarctic its birds and mammals have developed in a completely different way from those of the Arctic. Whereas seals are only relatively specialized in their food requirements, arctic birds of prey are very highly specialized in this respect. The long-tailed skua (Stercorarius longicaudus), for example, only breeds in regions where there are lemmings, although it may catch other small mammals in emergencies. This is apparently the situation an Spitsbergen, since it sometimes breeds near human settlements, where an occasional domestic mouse can be relied upon to venture into the open. It is an odd kind of specialization, considering that the skua spends the months when it is not breeding on the open ocean, feeding partly on dead animals and partly on food scrounged from other marine birds. The snowy owl (Nyctea) is

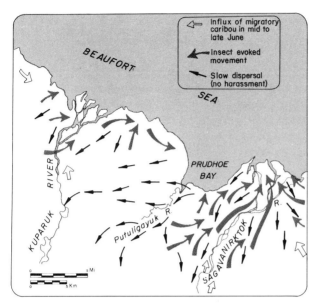

Fig. 55. An assessment of the effects of insect harassment on observed movements of caribou in the Prudhoe Bay area. (After White et al. in Brown, 1975)

also highly specialized on lemmings: Iceland and the Shetlands are the only places where it breeds outside the range of distribution of lemmings, and elsewhere its populations are closely coupled with the lemming cycle. As a rule it consumes no other food. This extremely narrow food specialization of the snowy owl may be due to something like imprinting. Snowy owls reared in zoos on white mice accept neither hamsters, lemmings nor brown mice even after fasting for several days. Live hamsters and lemmings evoke a panic reaction in the owl although they think nothing of catching living white mice or even large white rats. The third specialist in this series is the arctic gyr falcon (Falco rusticolus) whose occurrence is strictly coupled with the presence of ptarmigan (Lagopus lagopus and L. mutus), the two species usually accounting for more than 95% of its food. Its population numbers therefore fluctuate both regionally and temporally in very close dependence on the ptarmigan cycle.

Predators, above all of the flightless kind, undoubtedly play a very considerable role in the breeding distribution of warm-blooded arctic animals. The preference for small off-shore islands shown during the breeding season by ducks and geese, loons (Gavia), phalaropes (Phalaropus), gulls and terns is quite certainly connected with the polar fox (Alopex), which feeds largely on eggs and young birds at this time (Ryder, 1967). This explanation finds support in the fact that the only one flightless bird, the great auk, Pinguinus impennis, ever developed in the northern hemisphere and soon became extinct, although in the Antarctic, where terrestrial predators are completely lacking, the penguins have survived with great success.

Competition is almost entirely unknown in the Arctic. A possible case has been reported by Hinz (1976). The two most common groups of winged insects caught in formol traps in the Dryas tundra on Spitsbergen are mycetophilids and sciarids. Remarkably, however, a trap usually only contains members of one ot these two families, and comparable numbers of the two families have hardly ever been reported in one and the same trap. Mycetophilids are invariably later than sciarids, but as a rule a trap in which sciarids have been caught earlier does not contain mycetophilids at a later date, and vice versa. This was observed by Hinz in 1968 (1976), but had already been described in 1964 for other localities by Remmert, who also confirmed the findings at a different site, in 1973 and 1974. In any one series of traps neither of these authors was able to predict which trap would contain mycetophilids and which sciarids. This might be an example of strict interspecific competition between the two groups (i.e., between the predominant species, which is Exechia frigida in the case of the mycetophilids). In general, interspecific competition seems to be very slight, which is to be expected from a theoretical consideration of the ecological situation.

In spring when the snow begins to melt on the frozen ground the tundra is covered with pools, which are partly responsible for the plague of mosquitoes for which arctic regions are notorious. In the High Arctic, on Spitsbergen and Ellesmere Island, for example, the insects are far less troublesome. The mosquitoes (Aedes) seem to play a large role in the distribution of warm-blooded animals, and in the case of the reindeer this has long been an accepted fact even in the absence of reliable studies on the subject. The migration of reindeer to higher altitudes when the mosquitoes appear is suggestive. Careful investigations made in Prudhoe, Alaska (see p. 182) have unequivocally demonstrated the role played by culicids in the distribution of reindeer (Figs. 55, 126).

4. The Combination of Factors in the Arctic

Constant daylight in summer, constant darkness in winter, low temperatures combined with only small diurnal fluctuations and relatively low incoming radiation per unit of time (1 h or less); a minimum of shelter (arctic regions are north of the forest limit, and only in Scandinavia do they include any forests at all), and permafrost (again, only in Scandinavia is there little or none) are the factors which combine to constitute conditions which we recognize as being arctic. The whole is of such complexity that it has never been given adequate attention as such, so that the basis for discussion is very slender. A connection can be established between light and temperature, and its significance during the arctic winter and summer can be demonstrated. This was discussed by Wallgren (1954) in his comparison between the migratory ortolan bunting (Emberiza hortulana) and the non-migrant (or migrant only over short distances) gold hammer (E. citrinella). The somewhat smaller and more slender ortolan bunting is at a relative disadvantage with respect to heat budget: the little time available for feeding, together with the low temperatures of the winter months, make it impossible for it to survive the long winter nights. The gold hammer is better off and the short hours of daylight suffice, if enough food is available (Fig. 56).

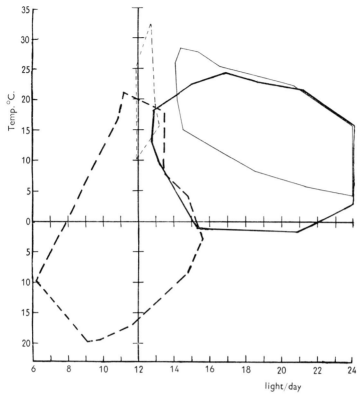

Fig. 56. Combined climographs. *Full thin line* ortolans during the summer. *Broken thin line* ortolans during the winter. *Full heavy line* yellow buntings during the summer. *Broken heavy line* yellow buntings during the winter. (Wallgren, 1954)

However, a nocturnal reduction in body temperature such as is known to take place in the tit, and which serves to reduce energy consumption during the cold hours of the night, can also be encountered in temperate continental regions with cold winters. Similarly, survival in deep snow, as described on p. 42ff., or the protection of plants by snow are not typical of the Arctic alone. Our knowledge of arctic animal ecology is thus more a matter of postulation than verified results.

The now well-known ecological laws were drawn up largely by Rensch (1936) and Mayr (1942; see Fisher, 1954 for a good, concise review). These laws can be formulated as follows for warm-blooded animals, i.e., for mammals and birds:

1. Allen's Law states that the protruding parts of the body such as tail, ears, beak, extremities etc. are relatively shorter in the cooler than in the warmer regions of the animal's area of distribution. The law can also usually be applied when making interspecific comparisons, e.g., a comparison between the red fox (Vulpes) and the polar fox (Alopex), or between a brown hare (Lepus europaeus) and a mountain hare (L. timidus).
2. Hesse's Law states that birds lay more eggs per batch in the cooler than in the warmer regions of their areas of distribution (Rensch, 1938).

3. Bergmann's Law stater that individuals in populations of a species nearer the pole tend to be larger and heavier than those of populations in warmer regions. This law is based on the principle that a relatively large body emits heat more slowly than a relatively small one of the same shape. This is also the explanation of Allen's Law.

Despite their very wide validity these laws are subject to recurrent attack. It cannot be denied that there are a large number of exceptions, but these are notably due to an unsatisfactory definition of the Arctic. With the definition employed here, both summer and winter temperatures are low and constant. If the definition is not strictly adhered to and continental regions south of the Arctic Circle are included, then very high summer temperatures and very low winter temperatures may be found, and the above laws will not apply. In particular, the laws connected with heat balance are only applicable where low temperatures prevail in both summer and winter. Hesse's Law is at present difficult to explain although Lack (1974) was able to confirm it. He correlated the size of a batch of eggs with the length of day and came to the conclusion that the number of eggs depends upon the amount of food that the parents can provide under the most favourable circumstances. Since these are a function of the hours of daylight the correlation would appear plausible at first sight. But in fact food density, the ease with which it can be found and the size of each individual food particle have also to be taken into consideration. The relatively large territories of arctic birds as compared to those of the same species in more southerly latitudes (Fig. 125), besides the fact that the numbers of both species and individuals of large poikilothermic animals decrease towards the pole, point to scarcity of food itself as a factor. It seems that the quantity of food that can possibly be collected per unit of time decreases gradually with increasing latitude, which undermines the plausibility of Lack's explanation of Hesse's Law.

Belopolski (1957) showed that there is a distinct reduction in the duration of the brooding and nestling stage of Alcidae in the North Atlantic with increasing latitude. Such a reduction is only found among typically arctic species but not among species that reach the northern limit of their area of distribution in boreal regions. So far there is no explanation for this phenomenon, which is not seen in land birds either (wagtail, starling), so that the earlier explanation that it is due to the fact that the birds can feed around the clock is no longer tenable.

Corresponding laws for terrestrial poikilothermic animals (successive decrease in herbivores and of relatively large forms) have already been discussed to some extent (see p. 32). Further adaptations of terrestrial poikilothermic animals to arctic conditions have been compiled by Oliver, 1958; Downes, 1965; and Corbet, 1971. Since the authors' definition of the Arctic does not coincide with our own the material is only partially applicable. Downes, for example, points out the increase in winglessness of arctic insects from Canada to the Aleutian Islands, although the "tundra" of the Aleutians cannot be regarded as arctic, but is a consequence of the strong wind prevailing on the islands. The authors stress the importance of temperature and sunshine in warming up arctic insects: the activity of arctic animals can cease abruptly if the sun becomes overcast. Among the simuliids the number of species that take up no food at the imagine stage increases towards the pole. In eight or nine genuinely arctic species of this family in Canada the adult female is a modified form that no longer feeds. Its mouth parts are too soft to pierce the skin and suck blood. In this way the especially perilous imagine stage is shortened and the increase in parthenogenesis in this group serves the same end. In Prosimulium ursinum in northern Norway most of the females do not even emerge from the pupa but die and disintegrate, the eggs landing in the rivers, where the young larvae hatch. Further to the south the imagines still hatch normally, but most populations in Norway are parthenogenetic.

In the light of more recent evidence the arguments of the two authors Downes and Corbet regarding diurnal rhythm and synchronization of the animals with the annual course of events must be considered obsolete. It now has to be assumed that all High-Arctic animals possess a physiological clock that is synchronized with the rotation of the earth. Many parameters that can readily be used as evidence for the existence of the physiological clock in southern latitudes are, however, useless in the High Arctic. In the latter regions parameters that exhibit a strict diurnal periodicity are often difficult to measure, since most organisms rest for only about one hour daily in the summer (see p. 9). In itself this is not a peculiarity of High-Arctic organisms but is predictable on the basis of the oscillation theory (see Daan and Aschoff, 1975). For adaptations of marine animals see p. 18.

The advantageous situation of warm-blooded animals as compared with poikilothermic organisms in the Arctic should not mislead us into assuming that the former have no difficulties in living there. The mortality rate, particularly in juvenile stages, is probably very high under the rigorous conditions of the Arctic. In general the choice of food available to terrestrial forms is very restricted. Under these circumstances a high rate of production is necessary, especially among smaller forms. Lemmings reproduce all year round, although the main breeding seasons are summer and winter, beneath the snow. On Devon Island the lowest numbers of lemmings are recorded in autumn, shortly before the beginning of winter reproduction. An adequate covering of snow is essential, otherwise the lemmings are unable to make good their continual losses. In its absence their numbers drop or the animals may even disappear altogether. Whether the same situation applies to the other small mammals of the High Arctic is at present uncertain.

The composition of the bird fauna of the High Arctic can be explained on a similar basis. There are fewer nidicolous species here than at other latitudes; the large majority are nidifugous, the young very soon being capable of fending for themselves. In view of the small amount of food available this means that the burden on the parent birds is lessened, the territories can be larger, and in some cases a second brood can even follow relatively soon, before the previous brood is capable of flight. The increase in size of territory with increasing latitude was clearly demonstrated by Holmes (1969) for the dunlin (Calidris alpina), whose territories at Point Barrow in the Arctic are about five times the size of those in the Subarctic at 61°N (Fig. 125).

A predisposition of many limicoles appears to have been exploited in adapting to the specific conditions of the Arctic. Often it is the male that chooses the nesting place and hatches the eggs alone whilst the female disappears from the breeding territory as soon as she has laid the eggs. Parmelee and Payne (1973) were able to show that the female sanderling (Crocethia alba) lays four eggs in each of two nests in rapid succession. The two nests may be quite far apart, and in this case female and male each brood one of the batches and rear the young alone. A very similar pattern of behaviour has been reported for Temminck's stint, Calidris temminckii (Hilden in Glutz, 1977). Here, too, the female leaves the first nest whilst the male remains to hatch and rear the brood. The female sets up a new territory with a second male and lays a second batch of eggs. As a rule, another female joins the first male, probably after she has already produced one batch of eggs with a different male. Whilst the male continues to brood, mating now begins once more, the female is fertilized and can rear a brood of her own. With the help of this system every individual adult bird of the species is in a position to hatch and rear a brood of its own.

A very highly specialized situation seems to exist in the case of the snow bunting, which is the northernmost passerine species. The observation of very recently fledged specimens near nests that were just being built suggest that several overlapping broods may be reared.

III. (Almost) Common Characteristics of Arctic Animals

1. Population Cycles

Ever since ancient times the regular occurrence of population cycles in the birds and mammals of the temperate, boreal and arctic zones of the Northern Hemisphere has attracted attention. The further north we proceed the more pronounced do such cycles become. In general three types can be distinguished:

1. Four-year cycles. Almost all small rodents of the Northern Hemisphere exhibit mass changes in a four-year cycle (Fig. 57). A characteristic feature is the simultaneous occurrence of the density maxima and minima in all species in a particular locality; e.g. lemmings (Lemmus) and voles (Clethrionomys, Microtus oeconomus and agrestis) achieve a maximum, or a minimum, at the same time.

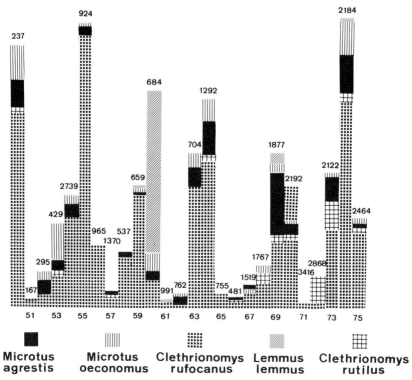

Fig. 57. Synchronous cycles: population cycles of small mammals in northern Finland. Data from captures made in the field. (After Lati from Remmert, 1980)

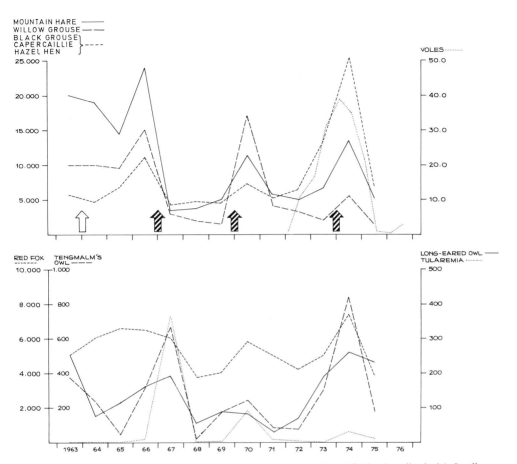

Fig. 58. Population fluctuations in voles (bank vole, grey-sided vole and field vole collectively). Small game (mountain hare, willow grouse, black grouse, capercaillie, hazel hen and red fox). Owls (Tengmalm's owl and long-eared owl) and tularemia as revealed by catches and literature survey, hunting statistics, bird ringing, and obligatory reporting of tularemia respectively. *Arrows* show winters with peak densities in voles whereas the *hatched* ones also indicate extensive forest damage (caused by voles) at the same time. (Hörnfeldt, 1978)

There is no confirmation of the repeated reports of wholesale migrations of lemmings at the peak of such a cycle, and in any case the densities involved seem to be too low (Fig. 60, 61).

In central Europe the long-tailed field mouse, the yellow-necked field mouse (Apodemus sylvaticus and A. flavicollis) and the common vole (M. arvalis) often exhibit parallel cycles. It is not known whether the same or comparable species (the snow vole, Microtus nivalis) undergo similar cycles in the central and south European mountains: if so, they must be very inconspicuous.

In parallel with these cycles there are also oscillations in the number of predators such as weasels (Mustela), foxes (Vulpes), owls (Striges), raptors and the long-tailed skua (Stercorarius longicaudus) that live on such animals. In Scandinavia hares

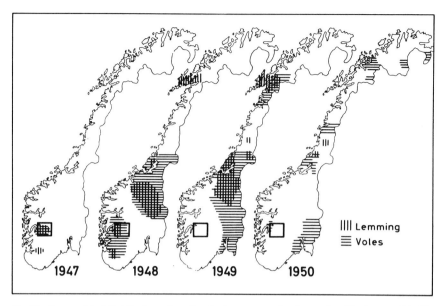

Fig. 59. Geographical distribution of small rodent occurrence (peak years) at Hardangervidda (*boxed*), as an example of regional variability of production of rodent biomass in Norway during a 4-year cycle. *Shaded areas* indicate mass occurrence of lemmings and other microtines in the years 1947-50

(Lepus) and tetraonid birds are also subject to a four-year cycle and Hörnfeldt (1978) has shown that their cycles are synchronized. In addition, the cycles of small rodents are obviously synchronized with those of the larger animals (Fig. 58). Although there are exceptions, synchronization seems to extend over very large areas. In general, the populations of large parts of the Scandinavian peninsula and Finland appear to fluctuate synchronously, although the amplitude varies from place to place (Fig. 59). Some passerine birds also appear to fit into these cycles (Fig. 60) other do not (Enemar et al., 1965).

An interesting fact that is seen again and again is that a species may exhibit very pronounced and regular cycles in population density in one part of its area of distribution but be relatively stable in another. For example, the well-known four-year lemming cycle at Point Barrow is entirely absent at Prudhoe, only 300 km away; and on Ellesmere Island the lemming populations do not show the typical oscillations seen on Devon Island, which in turn are not so distinct as those of Point Barrow. Another example is provided by the field vole (M. agrestis), which oscillates in north Sweden although there are non-oscillatory populations in South Sweden. There are indications that the ability to develop population cycles of this type is under genetic control (Nygren, 1978).

2. Cycles recurring approximately every nine years. These mainly involve the Tetraonidae of North America and are thus not synchronized with the small rodent populations fluctuating in four-year cycles (Figs. 61, 62). In addition, predators exhibit population cycles dependent upon cycles of the herbivores. This is best documented in the fur trade statistics for hares and lynx. A special situation is seen

Population Cycles

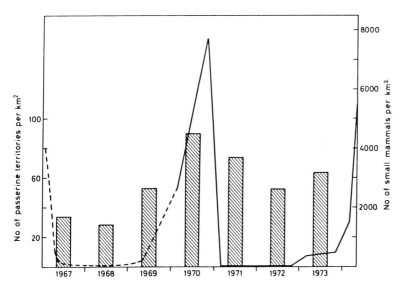

Fig. 60 Variations in the number of small mammals and passerine territories 1967–1973. Data from the Blåisen and Hansbu transects are used to illustrate bird territories. *Curve* represents small mammals, *hatched bars* passerine territories. (Lien in Wielgolaski, 1975)

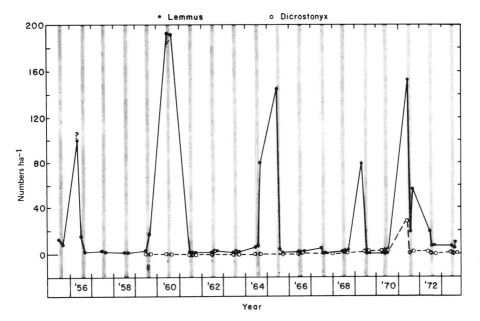

Fig. 61. Fluctuation of a lemming population at Point Barrow, Alaska. Note the relatively low number of lemmings per unit area at peaks of reproduction, with 1 animal per 50 m²

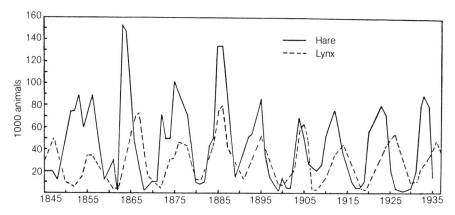

Fig. 62. Nine-year cycle of mountain hare and lynx in Canada

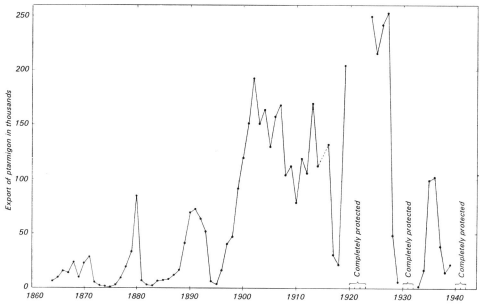

Fig. 63a. Oscillations in the annual export of Icelandic rock ptarmigan to Denmark. A nine-year cycle is faintly recognizable. (Gudmundson, 1960)

in Iceland, where the Iceland rock ptarmigan also appears to go through a nine-year cycle (Fig. 63a) whereas the long-tailed field mouse (Apodemus sylvaticus), which was certainly introduced by the Vikings from Scandinavia, exhibits no appreciable cycles.

Thus one and the same species of Lagopus oscillates in North America and Iceland in a nine-year cycle and in North Scandinavia in a four-year cycle. Although undoubtedly belonging to the same species as the Scandinavian ptarmigan, Lagopus in the alps shows no sign of a corresponding periodicity. In Scotland, on

Population Cycles

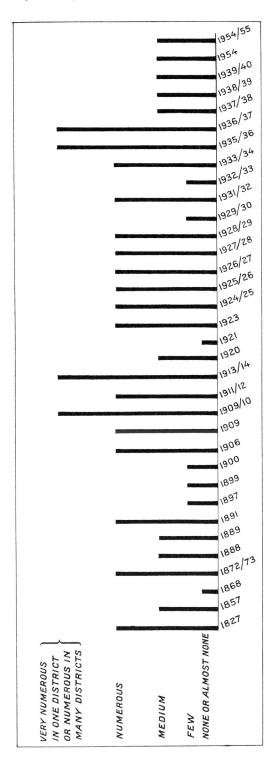

Fig. 63b. Fluctuations in the population of the Spitsbergen ptarmigan. (Løvenskiod, 1964)

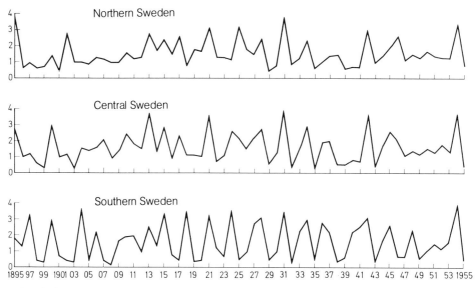

Fig. 64. Fluctuations in the seed production of spruce in northern Scandinavia. (After Svärdsson, 1957)

the other hand, grouse exhibit a four-year periodicity whilst on Spitsbergen, although there are very large fluctuations in the Lagopus population, there is no evidence of regular cycles (Fig. 63b).

3. There is some evidence that the caribou of North America and the reindeer of Spitsbergen fluctuate in a 70-year cycle. However, it is no simple matter to collect convincing evidence for a periodicity of this magnitude and it cannot yet be considered proved (see p. 201).

We now have to put the questions of cause and biological significance. The most obvious explanation of the oscillations in population density would be that they are dependent upon fluctuations in the food supplies. Such fluctuations are indeed familiar and there is an explanation for them (Svärdsson, 1957). Trees and shrubs for example do not produce seeds each year: following a seed year reserve materials have to be accumulated again and stored in the medullary rays. When sufficient quantities are available the appropriate environmental stimulus (which varies from species to species) can activate the energy produced in the green leaves and that stored in the medullary rays and divert it into the production of new seeds, which results in a regular cycle of seed production (Fig. 64). The cycles occur synchronously because the abiotic releasing factor appears to be the spring temperature —although the exact temperature varies from species to species. In a large-scale attempt Svärdsson tried to establish the connection between such "fat years" and the massive appearance of invading birds. The formation of seeds in Ericaceae also follows a cyclic pattern. In Scandinavia "seed years" often occur in a four-year cycle, so it is very likely that there is a connection with the animal cycles. A number of investigators have postulated a connection of this kind although as yet no unequivocal evidence has been offered. In fact, on the basis of their own investigations some authors (e.g., Tast and Kalela, 1971) have even denied that any such relationship exists. Another point to be considered is that vastly different individuals apparently turn up in the course of a cycle. Krebs and Myers succeeded in demonstrating genetic differences between the members of one and the same population of mice at its peak and its minimum. At the population peak selection favours those individuals that are capable of exceptionally aggressive defence of their food and personal territory, whereas at the population "low" individuals with a high rate of reproduction are at an advantage. Moss, Watson and Parr put forward a similar hypothesis for Lagopus, although not directly claiming that genetic factors are involved. They consider an overgrazing cycle of the essential parts of the food plants to play a governing role. In the meantime genetic differences have been found between Lagopus individuals from a dense population and those from a sparse population. The initial idea that predators play the main role in such cycles and that the sinusoid oscillations of prey and predator are 90° out of phase with each other has proved to be untenable. Nevertheless, the possibility cannot be excluded that at the peak of a cycle pathogenic organisms, parasites and even predators play a decisive role in bringing about the collapse of the population (see Hörnfeldt, 1978). The view is gradually gaining ground that population cycles are "initiated" by one or few species, and that other species are then drawn in. The importance of scarcity of food, deficiency in its quality, genetic changes in the course of a cycle, predators (especially in connection with the collapse of a population when predator stress suddenly becomes relatively strong), stress, pathogens and weather conditions varies from cycle to cycle, species to species and place to place (Fig. 65).

The significance of population oscillations is nowadays held to lie chiefly in the fact that they are unpredictable for predators. No predator or pathogenic organism is in a position to multiply in a comparably sudden manner. Constantly high population densities attract large numbers of predators, parasites and pathogens, but by oscillating it is possible for a species to maintain high mean population densities with comparatively little pressure from its enemies (Remmert, 1980).

The density of rodents exhibiting population cycles is usually overestimated for arctic latitudes. On Devon Island, for example, 2 or 3 individuals per hectare are reported for June, when numbers are at their highest (Bliss, 1977). In Prudhoe at a peak in population density 10–11 individuals/ha are reported (Feist in Brown, 1975), and at Point Barrow, the classical locality for lemming cycles, the figure of 200 individuals/ha of the brown lemming (Lemmus) is the highest ever reported at a peak, and the arctic lemming (Dicrostonyx) only rarely achieves a density of 20 individuals/ha. During a lemming "low" the density is 1–2 animals/ha. On the whole therefore, even at times of maximum increase a lemming individual has a territory of 50–100 m² at its disposal. Even under such conditions relatively high uric acid values could be demonstrated in the blood, indicative of severe density stress (Andrews et al., 1975).

On the basis of these data and from what is known about the food requirements and size of the territory of the snowy owl the following calculation can be made:

In summer one snowy owl requires 2–3 lemmings per day, i.e., a pair needs 5–6. The size of the breeding territory varies between 1 and 3 km² depending on the lemming density. The young owls are reared until capable of flight, during which

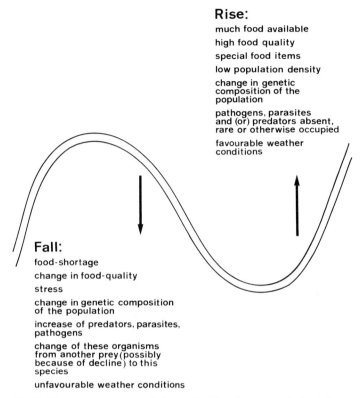

Fig. 65. Factors governing population cycles. The relative quantitative influence exerted by these factors may vary from cycle to cycle, from place to place and from species to species

time they consume about 150 lemmings. At times when lemming density is at a maximum as many as 9 young owls per brood can be reared, which accounts for 1300 lemmings. When the density is at a "low", however, there are usually only 300 lemmings per km² at the most, which is even less than the two parent owls require for themselves. But it cannot be expected that they meet with 100% success in their hunting activities, and in fact they are probably very unsuccessful at such low lemming densities. The available data suggest that prey species possessing a fixed territory, such as grouse, mice and related forms, are very elusive (summary in Remmert, 1978). It can be safely assumed that when the lemming density is low each animal will indeed possess a territory of its own which it knows so well that it is safe from the snowy owl. Under such conditions it is practically impossible for the latter to rear a brood. The situation is entirely different during the phase of mass reproduction, when 20,000 lemmings per km² can be reckoned with: 1300 of these would be needed by the owls for their young (assuming a brood of 9), plus 360 for the adults during the brooding and nestling period. In all, the owls would only eliminate about 1700 of the 20,000 individuals available per km² during a lemming peak.

In making this calculation we have ignored the fact that the lemmings reproduce 2–4 times annually, with an average litter of 4–6 young. Fewer offspring are produced during a "low" period, but the number rises as the population climbs to a maximum. Even on the basis of such rough estimates—and any attempt at more exact figures would only result in a deceptive impression of accuracy—it is clear that the snowy owl plays a negligible role at the peak of a lemming cycle, and that during its minimum only isolated vagrant birds can support themselves on chance catches but cannot risk breeding.

Obviously conclusions concerning the underlying causes of a particular cycle, however well it has been studied, cannot necessarily be applied to the next cycle of the same species in the same area, since different factors may be responsible for the decline and breakdown of the population. Still less can the conclusions be applied to cycles in other places, at different times and involving other species. Population oscillations of the kind discussed seem to be confined to the Northern Hemisphere and mainly to its boreal and arctic regions. Regular oscillations of a comparable nature have not been reported from the Southern Hemisphere, nor from mountainous regions.

What is the significance of these oscillations for arctic ecosystems?

A mass multiplication of the wild reindeer in southern Norway resulted in a serious depletion of the lichens which form their main source of food in winter. As a consequence the reindeer population underwent a partial collapse, and was still further reduced by excessive hunting. After a time the lichens began to grow once more and even ousted the vascular plants in some areas. A number of localities in the area are not frequented by the reindeer even at the height of their population peak. From a distance such areas have a pale yellowish-green appearance due to the large patches of reindeer lichens (Cladonia and Cetraria) covering the ground. Places occupied by the reindeer, however, usually have a thick carpet of phanerogams. Extrapolating these observations to regions where the reindeer populations may possibly oscillate in a 70-year cycle gives the following picture: as the numbers of reindeer rise the lichens are increasingly exploited to the point at which they are so seriously depleted that the reindeer population collapses for lack of winter food. After this, the lichens are able to recover slowly. Over-exploitation of the lichens does not result in bare patches of ground since vascular plants now fill up the spaces. This means, however, that in summer the warming influence of the sun cannot penetrate so deeply into the ground and it does not thaw to the same depth as before, so that the plants consequently receive less nutrients. Evidence that the cycle does in fact take the course described was found on Spitsbergen: in places with no reindeer extensive areas are covered with lichens, although there are scarcely any at all in localities grazed by the animals, whose density at present is very high (see Fig. 139). Instead a luxuriant vegetation consisting of many species of vascular plants is taking a hold. Should the reindeer population shrink the lichens would probably return and the ground once more thaw to a deeper level, so that the plants would again receive larger quantities of nutrients, probably including more phosphorous and other essential trace elements.

A cycle of this kind has been postulated by Piper and Schultz (Remmert, 1974) for the lemmings in Point Barrow, Alaska (Fig. 66), although analysis of the results does not seem to have progressed far enough for either confirmation or rejection of the hypothesis. In any case the oscillations of the mammals cannot be considered alone since they are certainly connected with alterations in the depth of

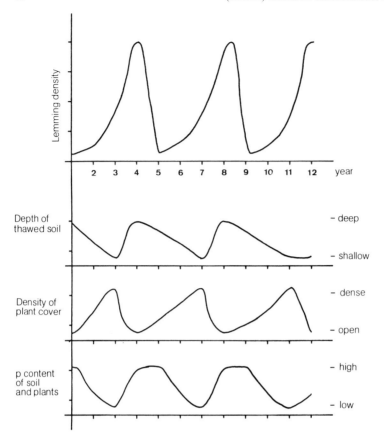

Fig. 66. The nutrient recovery cycle in Point Barrow, Alaska. (According to Piper and Schulz, after Remmert, 1974)

frozen ground and a very drastic change in the composition of the flora. An ecological study limited to a few years cannot provide an accurate picture of energy flow and nutrient cycling in an arctic ecosystem.

It might well be that the regularly occurring lemming cycles in some regions such as Devon Island are of particular importance for the decomposition of litter. In any case its breakdown is achieved more rapidly as a result of the lemming cycles than in their absence (Bliss, 1977).

Yet another consequence of these population cycles is the almost complete lack of habitual attachment to any one area shown by very many arctic raptors. Whereas in temperate latitudes it is usual for great owls (Bubo), peregrine falcons (Falco peregrinus) and gulls encountered in the breeding grounds to remain constant over decades or even centuries, the occurrence of the snowy owl (Nyctaea), gyr falcon (Falco rusticolus) and long-tailed skua (S. longicaudus) is practically unpredictable. The snowy owl and long-tailed skua and in many regions the pomarine skua (S. pomarinus) are highly specialized to hunting lemmings, seldom killing other species of vole, and very rarely grouse. Snowy owls may even be totally absent from the whole of Scandinavia for long periods of time, suddenly to

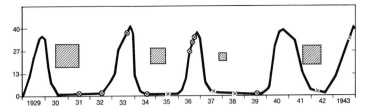

Fig. 67. Relation between population fluctuations of arctic lemmings in arctic Canada and the invasion of snowy owls in the east of the USA (*squares*). The size of these squares is proportional to the invasion by snowy owls. The owls go south when the number of lemmings is greatly reduced again after a peak. (Salomonsen, 1969)

reappear and breed at a time of mass increase in the number of lemmings. The gyr falcon (Falco rusticolus), however, is strictly specialized on Lagopus and its occurrence oscillates with the population of the latter. It is quite obvious that when there is a particularly large increase in lemming numbers owls and long-tailed skuas from a much larger region collect in the vicinity. The same applies to the gyr falcon in connection with Lagopus. It has not yet been possible to determine the extent of the regions from which the raptor populations are drawn. It seems likely that during population explosions of Lagopus in Iceland large numbers of the Greenland gyr falcon breed there. A still greater distance separates the breeding places of the snowy owl in Hardangervidda in southern Norway, only frequented in lemming years, and its breeding places in the Soviet Union, where it is encountered with relative consistency. What has been said here applies equally to other raptors such as the rough-legged buzzard (Buteo lagopus), the great grey owl (Strix nebulosa) and the hawk owl (Surnia ulula). They too may disappear for considerable periods of time from large tracts of country where they otherwise breed, such as the taiga regions of northern Sweden and northern Finland, suddenly to turn up again in quite large numbers.

It may be that the members of such species that are of circumpolar distribution constitute a single, large population. The large number of species with no racial subdivisions would support the idea, although, on the other hand, the proportion of white individuals of the gyr falcon in Greenland is conspicuously greater than in the European Arctic. There is clearly a pressing need for more exact information on the size of the reservoir from which a particular lemming area draws its raptors, as well as for studies concerning possible information systems able to function over very large distances.

Following a lemming year young raptors and owls are seen in relatively large numbers in southerly regions: young snowy owls turn up regularly as far away as the middle west of the United States of America (Fig. 67).

Mammalian predators, being less mobile than the birds, do not exhibit such large fluctuations, although their numbers do vary considerably according to the lemming numbers. Weasels (Mustela nivalis) in particular appear to play a role in controlling lemming populations, and in some cases they may increase as rapidly as the lemmings themselves. A computer model involving the data on lemming increase and collapse at Point Barrow only gave anything approaching satisfactory results if the weasels were also included.

2. Seasonal Migrations of Birds and Mammals

The majority of birds breeding in the Arctic and a large proportion of the mammals leave the true Arctic in winter for other regions. The following possibilities are open to them:

1. They may seek a locality closely resembling the environment in which they normally breed („ökologische Beharrungstendenz"). This is the alternative chosen by most migratory species.
2. Arctic birds, however, frequently prefer to seek an environment which is ecologically very different from that in which they breed. The bird's ecology usually "switches over" in its native locality at the moment when it feels the urge to migrate.
3. Some animals avoid only the worst of the winter by moving to regions where they can normally just survive the winter. Long distances are not covered (the distances and the part of the population involved vary very considerably according to the severity of the winter). This strategy is adopted by many of the mammals.
4. Finally there is the alternative of a seasonal change of biotope over shorter distances. This seems to be a widespread practice among small rodents.
5. The much discussed possibility, often attributed to the polar bear (Pedersen), of continual migration around the pole has not been confirmed. Polar bear populations, like those of other mammals, seem to be fairly firmly attached to a particular locality. The different populations are well separated from one another.

The „ökologische Beharrungstendenz" is exhibited at its best where a species remains as far north as is possible. But even then, radiation and day length are different from those in the breeding area. One possible way of maintaining almost identical conditions is by means of rapid migration to the Antarctic, which is what the Arctic tern does (Sterna paradisea).

Many arctic animals, especially marine birds, live on the pack ice of the Arctic Ocean or at the southern edge of the ice. (The birds often evade the winter by swimming southwards instead of migrating on the wing, returning north in the same way. This has also been observed in the Baltic Sea (Fig. 68).) The species chiefly concerned are the arctic Alcidae, High-Arctic gulls such as Pagophila eburnea and, to a lesser extent, Larus glaucoides, and in a modified form Rhodostetia rosea, a gull which migrates from its inland Siberian and Canadian breeding grounds to the frozen Arctic Ocean and probably spends the winter circling the pole.

The classical example of a bird that embarks upon long migrations to satisfy its need for similar environments both in winter and summer is the arctic tern (Sterna paradisea). It leaves its breeding grounds in early autumn and migrates relatively quickly and inconspicuously across the tropics to the Antarctic. The birds probably circle the antarctic continent in the wake of the winter storms (the southern summer) and then return to their northern breeding grounds (Salomonsen, 1967; Fig. 69). Even passerine species of the High Arctic seek ecologically similar regions

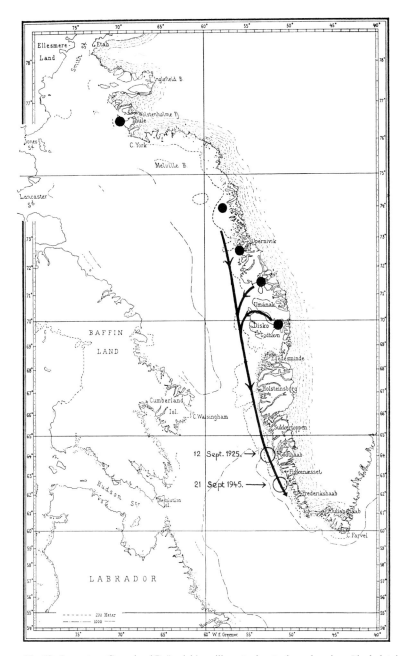

Fig. 68. In western Greenland Brünnich's guillemot migrates by swimming. *Black dots* indicate the most important breeding colonies, *thick lines* show the route taken by swimming birds over the fishing banks off the coast. *Open circles* with dates: observations of large flocks of swimming birds during migration. (After Salomonsen, 1969)

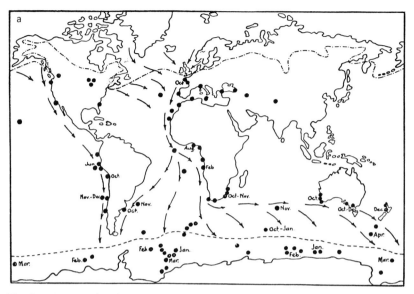

Fig. 69a. Autumn migration and eastward drift towards Australia and New Zealand of the Arctic Tern, indicated by *arrows*. *Solid circles* show records of the species outside of the breeding ground (not complete); –·–· southern limit of breeding range; - - - northern limit of Antarctic pack-ice belt. (After Salomonsen, 1967)

in winter. Snow buntings (Plectrophenax nivalis), Lapland buntings (Calcarius lapponicus) and the shore lark (Eremophila alpestris) overwinter in treeless localities such as the salt meadows of the Baltic coast of Germany.

A contrasting situation is seen in species, mainly birds, whose winter and summer requirements are entirely different. The long-tailed skua (S. longicaudus) is a specialized rodent predator in summer but spends the winter far out on the open sea, like Phalaropus species (Fig. 70) that in summer, in their breeding grounds, feed chiefly on Nematocera larvae. The phalaropes snap up small planktonic animals while the long-tailed skuas (and other species of skua) apparently obtain their food like the tropical frigate bird (Fregata) by stealing the prey of other birds. Sabine's gull, Larus (Xema) sabinii, although it exhibits no particular affinity for the sea in its breeding range, spends the winter in the productive regions off the coasts of West Africa and western South America. Loons, too, all of which breed on inland waters (the great northern loon, Gavia immer; the white-billed loon, G. adamsii; the black-throated loon, G. stallata), migrate to the open sea during the winter. After the young have become completely independent the birds are seldom seen in coastal areas.

The principle of biotope reversal is particularly well illustrated in the case of the purple sandpiper (Calidris maritima), which may breed in the open tundra or sometimes near the sea, preferably on sandy or silty beaches, sometimes in the vicinity of small pools, but often far from any kind of water. The birds appear to live chiefly on Nematocera larvae. Those that breed near the sea can often be seen, sometimes with their offspring in tow, on the muddy or sandy beaches searching for

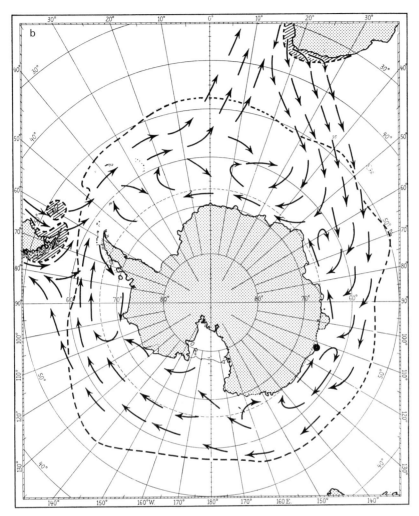

Fig. 69b. Probable route of the Arctic tern around the Antarctic continent in winter. The winter quarters off South America and South Africa are *hatched*, the position of the Antarctic recovery of a ringed Arctic tern is shown by the *solid circle*, the Antarctic Convergence by the *dashed line*. (After Salomonsen, 1967)

food. Quite abruptly the situation changes and small groups of the birds can be observed on rocks rising out of the sandy beaches, but no longer on the mud or sand beaches themselves, nor in the arctic inland regions. In its winter quarters the bird exclusively inhabits rocky coasts. The switch-over to this biotope is set off in the arctic breeding grounds, at the time of enhanced social attraction preceding migration.

The regular change of biotope as practised by many arctic birds (particularly sandpipers, phalaropes and gulls) presents the theoretical ecologist with problems hitherto overlooked. Theoretical ecology claims that the number of species in regions with a favourable climate is very much higher than in extreme regions and

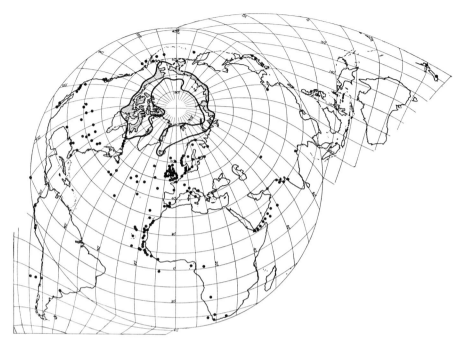

Fig. 70. Regions where the grey phalarope passes the winter. Breeding area embraced by *black line* and *shaded* overland. *Dots* represent some sight or specimen records. (After Fisher and Lockley, 1954)

hence the ecological factor governing the distribution of the species in the former is interspecific competition, whereas in regions with an unfavourable climate abiotic factors play the decisive role. This is all the more to be expected since regions with a good climate are of much greater geological age than those that form the cold regions of the earth today. For example, the community of organisms in tropical coral reefs has probably been in continuous existence longer than any other on the earth. For a new species to gain a foothold in a system of this kind, itself very rich in species on account of age and abiotic conditions, would seem to be impossible. But this is exactly what the arctic limicolae have apparently succeeded in doing. On the coral reefs of the Atlantic and Indian Oceans scarcely any indigenous species of birds seek food at ebb tide. (The only one is the rare Dromas ardeola, of the Indian Ocean.) Instead, it is almost exclusively species of the High Arctic such as the grey plover (Pluvialis squatarola), the whimbrel (Numenius phaeopus), the knot (Calidris canutus), and the sanderling (Crocethia alba) that exploit this reliable and abundant source of food. A similar state of affairs is seen on the shores of African inland waters where the sandpipers (wood sandpiper, Tringa glareola; green sandpiper, T. ochropus, marsh sandpiper, T. stagnatilis, common sandpiper, Actitis hypoleucos etc.) of the boreal zone are practically without competition for the food at hand. This is a problem that has so far been ignored by theoretical ecologists. It is very odd that this long-existent and rich supply of food should not have been exploited by a species breeding in the same geographical region, and that no indigenous bird species has evolved in this direction, particularly in view of

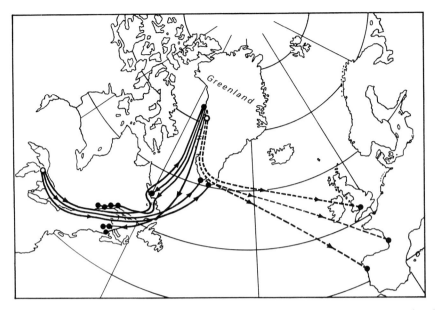

Fig. 71. *Broken lines* indicate the routes taken by the Greenland wheatear in its autumn migration to Europe (whence it emigrated to Greenland). *Continuous lines* depict the routes taken by the snow bunting to North America. (After Salomonsen, 1969)

presumably severe pressure from interspecific competition (which has in some cases been confirmed).

There is still another question that has so far been ignored by physiologists. How do the small species of birds that breed in the extreme High Arctic manage to spend the winter in regions with the strongest sunshine and very high temperatures? Most mammals would have considerable difficulty in coping with such enormous differences in radiation and temperature. These birds apparently have none.

The routes taken by migrating birds can in part be traced back to the original routes followed by the birds in the course of their distribution. The wheatear (Oenanthe) of Greenland migrates via Europe to the south, the snow bunting (Plectrophenax) via the Arctic Islands to North America (Fig. 71). However, the explanation is partly of an ecological nature. For example, the black-throated divers of northern Siberia (Gavia arctica) migrate across the continent of Asia towards the south by crossing the Caspian Sea to the Black Sea, whereas the northerly migration route crosses Europe to the Baltic Sea and thence to northern Siberia. The reasons are obvious. During the autumn migration the Soviet Union is still relatively warm and the routes to the south are open. But when the birds are ready to return to the north to start breeding the autumn route is still frozen. At this time, however, the route to the Baltic and thence to the north is open and probably the Arctic Ocean as well. The birds can then wait in the immediate vicinity of their breeding grounds until the onset of spring when they begin right away with the business of breeding (Fig. 72). The American golden plover (Pluvialis domenica) does something similar. In late autumn it migrates from its breeding grounds down

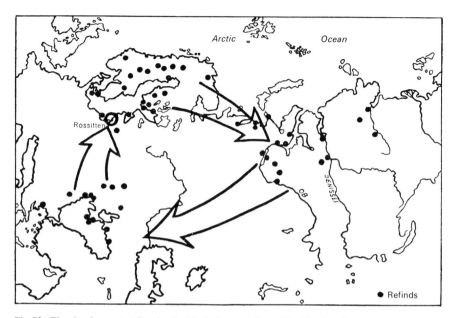

Fig. 72. The circular route taken by the black-throated diver. The Siberian diver takes a direct route to its winter quarters on the Black Sea. The homeward journey to the breeding grounds between Ob and Lena involves a detour across East Prussia and the Baltic Sea. (After Schüz, 1952)

the North American coast, across the western Atlantic to South America (Fig. 73), but returns along the inland route. The coastal detour is now unnecessary since the continent of North America warms up at much the same rate from south to north. Furthermore, towards the middle or end of May a sudden surge of warm air regularly rushes along the vast Mississippi river valley almost up to the Canadian border, warming the inland regions. Enormous numbers of passerines (for example the American humming-bird) accompany the warm air masses far to the north. This route is also taken by the golden plover, which can thus reach its breeding grounds far more economically and sooner than by following the sea route. In some years the ice between the islands of the Canadian Archipelago does not melt at all, even when the tundra has completely warmed up and provides the golden plover with ideal conditions for breeding.

In extreme arctic regions animals that are generally held to adhere to one locality may also exhibit annual migratory rhythms. A considerable proportion of the willow ptarmigan (Lagopus lagopus) of northern Alaska migrates south in winter. Up to 10,000 individuals daily have been observed in October on the Anaktuvuk Pass. The return flight begins in January and continues into May. In all, 50,000 individuals are in transit every autumn and spring. Older males tend to migrate less than young animals and females (Irving et al., 1967).

Most arctic mammals remain in the same geographical region winter and summer. The muskox (Ovibos moschatus) is a good example. Small rodents frequently undertake minor migrations, usually to a different biotope. Lemmings, for example, alternate in an annual cycle between wet and dry areas of the tundra.

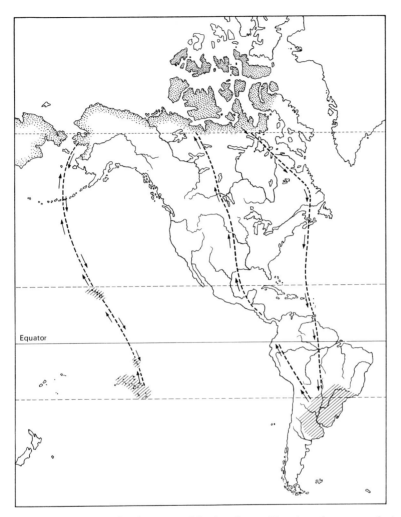

Fig. 73. The circular migration route of the American golden plover between arctic America and the Pampas of South America, and the route from the Bering Straits to the South Sea Islands taken by the east Siberian golden plover. (After Schüz, 1952)

The wet areas offer them food, mainly mosses, in summer, whereas the dry areas are better suited for the winter, which they mainly spend beneath the snow, feeding on mosses and the growing tips of grasses (Kalela, 1971).

Many populations of the North American caribou and of the old-world reindeer undertake migrations over what are for terrestrial mammals very long distances (Fig. 74). In autumn they wander in large herds many hundreds of kilometres to the south and return along approximately the same routes in spring to their summer grazing grounds. Caribou and reindeer may also move from the tundra into the forests in winter, in which case their food is limited to tree mosses if the snow is deep. Not all reindeer populations migrate, an exception being the wild

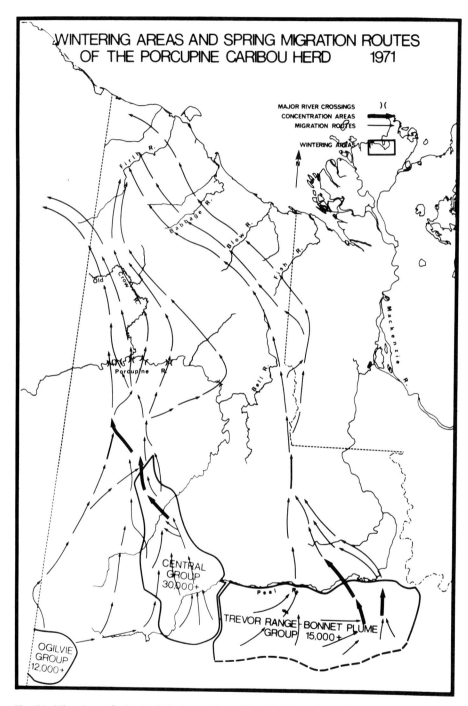

Fig. 74. Migrations of a herd of Alaskan caribou. (From Luick et al., 1975)

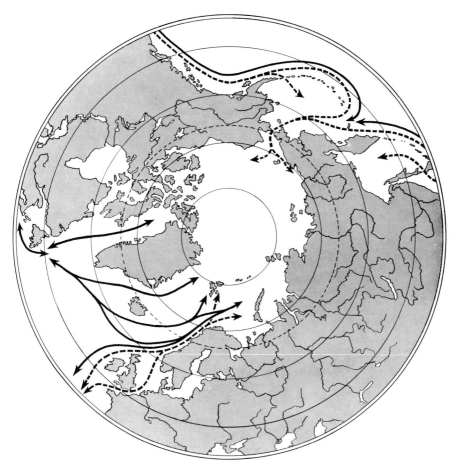

Fig. 75. Migratory routes of the blue whale (— —) and the fin whale (– – –) in the Northern Hemisphere. (After Slipjer, 1958 from Gerlach, 1965)

reindeer of southern Norway. The platyrhynchos race on Spitsbergen and similar forms on Peary Island and Ellesmere Island remain in their native areas summer and winter. Thus the most northern populations of reindeer and caribou do not migrate. Systematic problems arise in connection with the southerly migrations of the continental forms. The forest forms that live to the south of the wide variety of tundra forms also differ considerably from one another. The antlers of the forest forms are relatively thick and less spreading, more like those of the red deer. The animals are altogether stockier and not so obviously suited to swift movement as the tundra forms. In contrast to the almost invariably large herds of tundra forms, the forest forms live alone or in small groups. Despite the fact that the two can sometimes be encountered in the same area in winter there are apparently no bastards (domestic reindeer are descended from the tundra reindeer and their penetration of the forest regions of Scandinavia is due to anthropogenic factors).

A number of marine mammals such as the whales, (Fig. 75) and some seals that bear their young on the pack ice in spring (Fig. 76), migrate over considerable

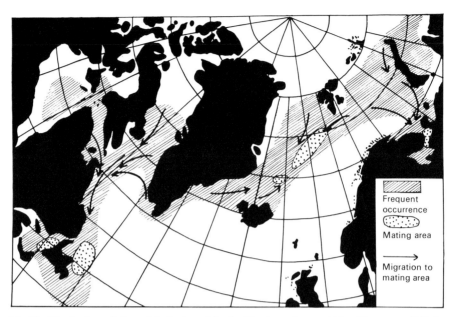

Fig. 76. Distribution of the saddle-back seal in the Northern Ocean. (After Sievertsen, 1941 from Gerlach, 1965)

distances. Only the narwhale (Monodon) and the white whale (Delphinapterus) remain in the High Arctic all the year round.

Teleosts are thought to migrate, but our knowledge on the subject is very limited. The tunny travels northwards in summer, almost up to the Arctic, returning south in winter to tropical regions.

No one explanation can account for all of these migrations. In part, they are a result of scarcity of food, in part they are a response to a combination of cold and darkness (or less daylight), as Wallgren (1954) was able to show by comparing the gold hammer and the ortolan (see p. 66). Meinertzhagen has drawn attention to the fact that in many cases scarcity of food cannot play a decisive role since adequate supplies are available, which is the case in most years for the passerine species of the taiga and birch forests. The same author also pointed out that passerines require grit for grinding their food in the gizzard, so that in regions with a very deep blanket of snow this may represent a decisive factor: the stones are usually worn down within about a week. This theory finds support in the ever-increasing tendency for birds to overwinter in the vicinity of human dwellings and along many roads that are strewn with gravel in winter, i.e., near a regular supply of gizzard stones.

All larger migrations are of quite considerable importance for the tundra ecosystem. Food is not available in the same quantities all the year round (Fig. 77). Because herbivorous species are not in the tundra at its most sensitive time, which is the very early spring, but are seeking food further south, the region is able to support a remarkably large number of herbivores at other times without any difficulty. The red deer (Cervus) of temperate latitudes also used to migrate in just the same way as the tundra reindeer nowadays. Large herds formed in autumn and

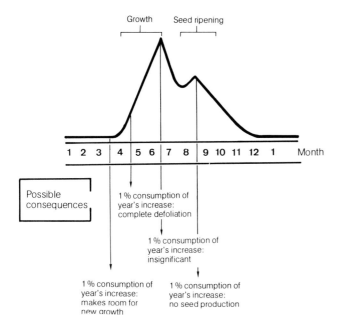

Fig. 77. The importance of warm-blooded herbivores for the plants in their environment. (After Remmert, 1980)

spent the winter feeding on the soft woody plants in the vicinity of rivers, not returning to their mountainous quarters until long after the plants there had begun to sprout. Now that migrations of this kind have become impossible due to the increased density of the human population, considerable damage can be afflicted on the vegetation by even a few red deer. A similar situation is seen in connection with larger herbivorous birds. For example, the high density of wild geese supported by "oases" in central Iceland (see Gardasson) is only possible because the birds do not return to these breeding grounds until the most sensitive phase is already past. This is the only way of maintaining the practically unequalled rate of utilization of the primary production over longer periods of time: in fact almost 100% of the primary production is consumed by the geese. Non-migrant herbivorous populations are either of low density or their numbers exhibit very strong fluctuations.

3. Entrainment of Animals to the Yearly Cycle

This is a topic that can only be treated cursorily since we still know far too little about what goes on during the darkness of winter. As a rule the mating season of the large mammals is in autumn so that they bear their young in spring or late winter, which are favourable times. Two examples will be described in some detail in order to illustrate the complications and consequences of this mode of behaviour. Firstly, it places a very heavy burden upon the adult animal, since the females are carrying their young during the most difficult part of the winter. It seems as if the development of antlers in caribou and reindeer entirely serves the ends of this reproductive cycle. It is striking that only under the particularly difficult conditions of the Arctic have the females developed antlers, in contrast to the cervids in temperate and tropical zones. Such an inconsistency is only explicable if seen in connection with the reproductive cycle. It is the animal with the strongest antlers that dominates the group and enjoys priority with respect to food. The antlers of the males begin to grow in spring before those of the females, but they are cast at the end of the rutting season in autumn, before the actual winter begins. The antlers of the non-pregnant females begin to grow later, and are not discarded until the end of the year. Thus it is these females that are dominant during the first part of the winter: they can hold their own against the males and secure enough food, which is very important since most of them are by this time pregnant. The antlers of the mother animals begin to grow much later, long after the birth of their young, and are still barely detectable when those of the males are fully developed. These females shed their relatively small antlers only shortly before the birth of the next calf. In this way a clear dominance is secured for the time, at the end of winter, when food is at its scarcest and the mother animals, which were not so well able to build up reserves in the autumn, are particularly important for the calves. The apparent paradox that under the severe conditions of the Arctic the females allow themselves the luxury of antlers can probably be regarded as a strategy for overcoming unfavourable environmental conditions and for ensuring success in intraspecific competition.

It is interesting that the largest penguin of all (Aptenodytes forsteri), which is incapable of flight and biologically resembles a seal more than a bird, has evolved a similar reproductive cycle. Like the arctic mammals it, too, reproduces in winter. When, as winter approaches, all other antarctic species move northwards and nearer to the open ocean where it is not quite so cold and where food is always available, the penguins remain where they are or even move slightly further south. In mid-winter, in the depths of the polar darkness and when temperatures are so low that all other warm-blooded animals either leave the continent altogether or, like the Weddell seal, take to the water, the emperor penguin begins to breed. The southernmost colony, consisting of up to 20,000 individuals, is in the midst of the pack ice of the Ross Sea near Cape Crozier, about 200 km west of McMurdo station at almost 80°S. Once pairs have been formed a lifelong partnership commences with a complicated mating ritual. The egg is laid on the firm ice, which is not so cold as the land because the water, which only has a temperature of $-1.7°$ C, warms it up slightly. The egg weighs about 450 g, is 12 cm long and measures 8 cm across. The male at once takes charge, rolls it on to its feet with the help of its beak and covers it with a finely feathered abdominal fold. For 60 days the male remains on the egg in a fasting condition, despite darkness and cold. If the temperature drops to below $-30°$ C, for example, the animal has to maintain a temperature gradient of at least $60°$ C with the help of feet and abdominal fold, because the temperature of the egg must not drop below $30°$ C. However, mere figures tell us little. They tell us nothing of the storms with wind velocities of up to

130 km/h (Table 3), whose cooling effect is much greater than normal air movements because of the accompanying ice crystals. At an air temperature of $-20°$ C a blizzard with a velocity of 40 m/s deprives the body of as much heat as stationary air at $-180°$ C. Thus the ice storms of the Antarctic represent the most extreme conditions of cold to which warm-blooded animals are ever exposed. In very intense cold the penguins huddle together to form a compact mass in which one body warms another. The female in the meantime migrates 100–200 km northwards across the ice until it reaches open water and food. The young hatch after 2 months, during which time or even longer the male has had no food whatever. It now feeds the young bird with a whitish, highly nutrient fluid secreted in its crop, and which for the next 4–5 weeks is often the only food that the young birds receive. From the time when they begin to brood until the end of this feeding period the weight of the males drops from 37 to about 25 kg. The females now return, fortified by their food reserves, and take over the job of feeding the young whilst the males wander northwards to the edge of the ice to feed. In the meantime the arctic spring has set in, the edge of the ice approaches the colonies and the young can at last become independent. The adults now begin to moult, which was impossible during the winter.

In the case of seals the birth of the young and mating are combined, and as soon as the pups have been born the next mating season commences. The gestation period is 11 months in all seals, irrespective of their size (Table 4). Only in this way is the pelagic way of life of many seals at all possible: they climb on to the ice merely to reproduce and spend the rest of the time in the open sea, independent of land and ice.

The behaviour of whales is heterogeneous. The large species migrate from arctic waters to tropical or at least subtropical regions to breed (as is shown in Fig. 78) for the fin whale). The young are thus born in relatively warm oceanic regions, and during these months mating takes place, before the animals return to their arctic or antarctic feeding grounds. At the end of the summer they once more head for tropical waters to bear their young. The very typical arctic species such as the narwhale (Monodon) and the white whale (Delphinapterus) behave differently. The former is said to be able to mate throughout the year and accordingly to bear its young at any time, whereas the mating season of the white whale is in late summer and autumn, its young being born in June/July. It is probably comparable in this respect to terrestrial mammals.

A completely different pattern of behaviour is seen in small terrestrial mammals. They invariably breed in summer, when several batches of young may be born in succession. To what extent winter breeding is possible in lemmings is uncertain. They often exhibit an annual cycle of migration between different habitats but they reproduce in both their summer and winter territories, in winter under a blanket of snow.

Reproduction in all birds, except the antarctic emperor penguin as described above, is limited to the summer months. The young therefore hatch when food supplies are at their best. Nettleship (1974) reported that the hatching of the majority of successful broods on Ellesmere Island coincided with the peak in hatching of chironomids.

Table 4. Biological data on arctic mammals. (Van den Brink, 1957)

Species name	Mating period	Gestation period	Time of birth	Annual number of litters	Size of litter	Young independent after	Sexual maturity at (in year)	Duration of life (in years)
Wolverine	IV–V	8–9 month	I–II	1	2–4 (1–5)	$\frac{1}{2}$–$\frac{3}{4}$ year	$2\frac{1}{4}$	15
Walrus	VI–VII	11 month	V–VI	1	1	$1\frac{1}{2}$–2 year	4–5	20–30
Harbour seal	VII–VIII	11 month	VI–VII	1	1–2	8 weeks	3	15–20
Ringed seal	V–VI	11 month	IV–V	1	1 (2)	6 weeks	3–4	15–20?
Saddle-back seal	III–VI	11 month	II–V	1	1	2 weeks	3–4	20–30?
Bearded seal	VIII–IX	11 month	VI–VII	1	1	6 weeks	3–4	20–30?
Atlantic seal	XI–III	11 month	X–II	1	1	6 weeks	3–4	40?
Hooded seal	IV–V	11 month	III–IV	1	1	5 weeks	3–4	20–30?
Reindeer	IX–X	32 weeks	V–VI	1	2 (1)	$\frac{3}{4}$–1 year	$1\frac{1}{2}$	15
Muskox	VII–IX	34–36 weeks	IV–V	1	1–2	$1\frac{1}{2}$–2 year	3–4	20–25
White whale	VI–X	10 month	VI–VII	1	1	1 year	1?	?
Narwhale	all year	10–12 month?	all year	1		1 year	1?	?
Grey whale	XI–V	12 month	XI–V	1	1–2	2 year	2?	?
Fin whale	XI–III N. Arctic Ocean V–IX S. Arctic Ocean	$11\frac{1}{2}$ month	X–II N. Arctic Ocean IV–IX S. Arctic Ocean	1	1 (2–3)	2 year	2	30–35?
Sei whale	XI–III N. Arctic Ocean V–IX S. Arctic Ocean	$12\frac{1}{2}$ month	XII–IV N. Arctic Ocean V–X S. Arctic Ocean	1	1 (2)	$1\frac{1}{2}$ year	$1\frac{1}{2}$	25–30?
Minke whale	I–V	10 month	IX–III	1	1 (2)	10–12 month?	2	25–30?
Blue whale	XI–III N. Arctic Ocean V–IX S. Arctic Ocean	$10\frac{1}{2}$ month	IX–I N. Arctic Ocean III–VII S. Arctic Ocean	1	1 (2)	$1\frac{3}{4}$ year	2	35–40
Humpback whale	III–IV N. Arctic Ocean IX S. Arctic Ocean	11 month	II–III N. Arctic Ocean X S. Arctic Ocean	1	1 (2)	10–12 month?	22 month	30–35?
N. Atlantic right whale	XXI–IV	9–10 month	X–II	1	1 (2?)	$1\frac{3}{4}$ year	2	30–40?
Greenland right whale	VII–VIII	9–10 month	III–IV	1	1–2	$1\frac{3}{4}$ year	2	30–40?

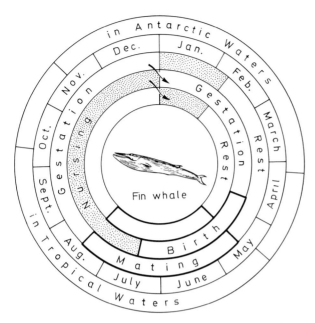

Fig. 78. Annual cycle of the fin whale in the Antarctic. In the Arctic the yearly cycle is displaced by 6 months. (After Slijper, 1962)

It is nevertheless unrealistic to limit the discussion of birds to their breeding habits. The fulmar (Fulmarus glacialis) provides us with a good illustration of this fact. As Fisher (1952) showed, the birds visit the large and familiar cliffs in the winter, after which some of them disappear again until breeding commences at the end of May to mid-June (Fig. 79). The same can also be observed at stations in the High Arctic: on Jan Mayen, on Bear Island and on Spitsbergen for example, the first birds appear on the breeding cliffs from mid-December until mid-January, although the eggs are not laid before the end of May to mid-June. Only the extreme frozen regions such as Greenland, the Canadian Archipelago and Franz Josef's land are not visited by birds before mid-March. Breeding begins here, too, at the end of May to mid-June.

We know that many birds and mammals of temperate latitudes exhibit quantitative as well as qualitative differences in food uptake from one time of year to another. A famous example is that of the roe deer (Capreolus capreolus), about which much has been learned from Ellenberg's (1978) studies. In winter it requires amazingly little food, whereas it consumes very large quantities at the height of summer (when it is said to be "in grease") and consequently the weight of the animals fluctuates during the course of the year. A similar situation has been described for the capercaillie (Tetrao urogallus), whose weight maximum is achieved in November. However, the annual weight cycle is partially masked within the population by weight differences due to age, sex and temperature. Premature cold in autumn, for example, can cause the weight maximum to be attained earlier.

The behaviour of the animals is in fact exactly the opposite of what would be expected: instead of taking up larger quantities of food in winter to compensate for

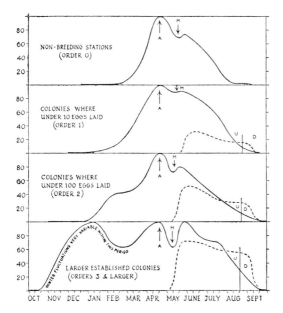

Fig. 79. Diagram of the population of fulmars throughout the year in colonies of different sizes in Britain; represented in percentage of the peak population, which is almost always in late April. *Thick line* adults; *broken line* eggs or young; *A* April peak; *M* May dip; *U* always some unattended young; *D* young deserted by parents. (Fisher, 1952)

the greater heat losses they do just the reverse. It should be added that no really good long-term studies have so far been carried out on the true arctic animal species with regard to seasonal variations in food intake and body weight. Ceska (1974) demonstrated that in central European zoos snowy owls also eat less food in winter than in autumn (Table 5). Feeding experiments involving the arctic fox under arctic conditions on Devon Island revealed the extremely high food consumption in summer of 370 kcal per kg body weight and day as opposed to merely 63 kcal per kg and day in winter (Bliss, 1977; Table 6). The white whale in the Duisburg Zoo behaved rather differently: whereas in August it consumed only 6 kg of fish daily its consumption reached a maximum of 12 kg per day in April, May and June. In the winter months of December, January and February intermediate values of 7–9 kg

Table 5. A comparison between the food consumption of free-living snowy owls and those in the Nürnberg Zoo (NZ). (Ceska, 1974)

	Alaska	NZ
Food intake 1 adult animal/year in kg live wt.	600–1600 lemmings $\varnothing$1 lemming à 80 g $\cong$ 55–130 kg	2760 mice $\varnothing$1 mouse à 25 g $\cong$ 69 kg
Food intake 1 adult animal/year in g live wt.	Estimated 150–350 g	$\varnothing$219 g
Food requirements of young during rearing in kg live wt.	1300 lemmings $\varnothing$1 lemming à 80 g $\cong$ 100 kg (9 young)	2360 mice $\varnothing$1 mouse à 25 g $\cong$ 59 kg (5 young)
Food intake of a young animal/day during rearing in g live wt.	160 g $\cong$ 2 lemmings à 80 g	132 g $\cong$ 5.6 mice à 25 g

Table 6. Food requirements and ecological efficiency of terrestrial mammals and birds of the Arctic (demand/day)

Species	Live weight (kg)	Consumption (kcal)	Daily faeces + urine (kcal)	Assimilation (% A)	Calculated BMR/day (kcal)	Remarks
Arctic hare	3.87	1,696.2	834.1	50.8	193	Bliss (1977), field experiments
Muskox	259	4,907	1,380	71.8	4,519	Bliss (1977), field experiments
Reindeer, zoo	55	6,400	900	86	1,400	Pöhlmann (1976), extra zoo food
Reindeer, Spitsbergen	55	11,000	3,260	70	1,400	Pöhlmann (1976), calculated from droppings in the field and from N-budget
Lemming dicrostonyx	0.047	88.3	21.5	77	7.0	Bliss (1977)
Willow ptarmigan	0.45	171	45.7	70	39	West (1972); extra zoo food; assimilation efficiency lower under field conditions
Goose (pink-foot)	4	965	227	76.4	200	Pöhlmann (1976), extra zoo food
Snowy owl	1.97	50 g dry wt. 85 g dry wt. late summer			137	Ceska (1978), zoo and field
Weasel	0.1	15.36 g dry wt.		73.9	12.4	Bliss (1977), feeding in the field
Fox (arctic)	1.3	205.9 dry wt.		86.3	85	Bliss (1977), feeding in the field [a]

[a] Very large seasonal differences: 370 kcal per kg body weight and day in summer, 63 kcal in winter

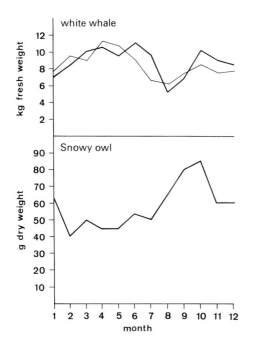

Fig. 80. Daily food intake by the white whale (Delphinapterus leucas, *above*) and the snowy owl (Nyctaea scandiaca, *below*) over the course of the year in German zoological gardens. (After Gewalt and Ceska, 1976)

per animal and day were reported for the two individuals studied (Gewalt, 1976; Fig. 80). From a number of other observations it seems fairly certain that reindeer and muskoxen also eat relatively little in winter, drawing upon the fat reserves laid up in summer. Apparently the elevation of basic metabolism due to severe

Table 7. Daily food requirements of marine birds and mammals in zoological gardens. The food of all species consists entirely of fish

Species	Weight (kg)	Food per day (kg)	= kcal/day	BMR (kcal/day [b])
Uria aalge	1	0,5	500	89
Spheniscus humboldti	4.5	1.2	1,250	267
Eudyptes crestatus	4	1	1,000	244
Aptenodytes patagonicus	20	2.5	2,500	800
Aptenodytes patagonicus	14	1.4	1,400	600
Aptenodytes patagonicus [a]	11.5	4	4,000	530
Phoca vitulina	50	1.6	1,600	1,300
Callorhinus ursinus	35	3.5	3,500	1,000
Calophus californianus	150	5.5	5,500	3,000
Calophus californianus	270	8.5	8,500	4,700
Otaria sp.	150	5.6	5,600	3,000
Otaria sp.	180	8.5	8,500	3,500
Eumetopias jubatus	625	13	13,000	9,750
Tursiops truncatus	180	5.5	5,500	3,500
Delphinapterus leucas	900–1,200	8–15	∅11,500	12,900

[a] Kept at temperatures around freezing point
[b] Calculated from the formula kcal/day = $89 \times$ body weight$^{0.73}$ (birds), $70 \times$ body weight$^{0.75}$ (mammals)

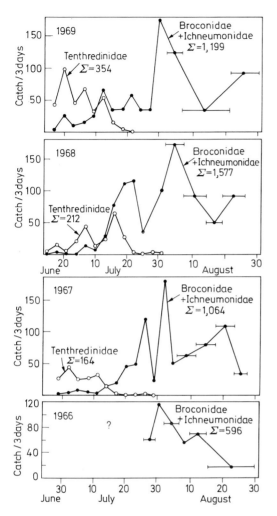

Fig. 81a. Occurrence of herbivorous sawflies and parasitic Hymenoptera in pitfall traps near Barrow, Alaska. (After McLean and Pitelka, 1971)

cold need not be reflected in increased food intake. Food requirements and energy consumption may be temporarily divergent. Since most of the data in Table 7 were obtained from short-term feeding experiments they should be regarded with a certain amount of scepticism, particularly those concerning the food requirements of penguins in zoological gardens. The values from the exotarium of the Frankfurt Zoo for king penguins kept at constant temperatures around freezing-point are much higher than those for the same species in other zoos where the birds are kept under central European conditions. This is probably the explanation for the discrepancies between the calculated values for basic metabolism and the quantity of food actually consumed.

The way in which nature completes its annual course in the short arctic summer can be illustrated by a brief outline of the situation pertaining to insects. The frequently encountered statement that the arctic summer is an undifferentiated

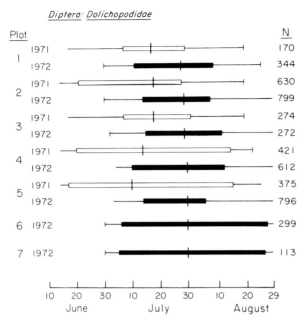

Fig. 81b. Seasonal distribution of "sticky-board" captures of long-legged flies (Dolichopodidae) in 1971 and 1972 at Prudhoe, Alaska. (After McLean in Brown, 1975)

season does not even hold for the High Arctic regions with a growing season of little more than a month. This has been clearly shown by studies carried out in Point Barrow, Prudhoe and central Spitsbergen (MacLean and Pitelka, 1971; Brown, 1975; Hinz, 1976; Remmert, unpublished). On Spitsbergen the sciarids can be considered "spring" animals, appearing in the first half of July, whilst the mycetophilids, parasitic Hymenoptera and flies represent the mid- or late-summer animals. The terms mid- or late-summer species apply particularly to aphids, empidids (Rhamphomyia caudata) and syrphids (Epistrophe tarsata). The situation in Point Barrow is in principle the same. Here, too, the first animals to appear are the very small Nematocera, followed later by the larger ones, whilst the parasitic Hymenoptera are the typical species of summer and autumn. In Prudhoe, too, much the same was found, with small Nematocera appearing relatively early, followed later by parasitic Hymenoptera and cicadas (Fig. 81). Considerable differences in timing, however, occur from one year to the next, so that the flight of tipulids in Prudhoe, for example, did not overlap at all in 1971 and 1972. Similar discrepancies were seen on Spitsbergen. There are also comparable differences in the times at which plants come into bloom. For example, on Spitsbergen Saxifraga hirculus flowers later than most of the other Saxifraga species.

The zeitgeber responsible for the entrainment of organisms to the arctic seasons has formed the subject of much speculation. The fact that in the High Arctic, for example on Spitsbergen or Ellesmere Island, it is light throughout the whole of the growing season from the beginning of April until the beginning of August, has led to the widespread assumption that photoperiod is not a controlling factor. The results obtained by Stross and Kangas (1974) on Daphnia middendorffia, and the studies of Vaartaja on North American trees contradict this view. Daphnia in the Arctic produce a brood in spring in small bodies of water and lakes, after which both they and their offspring produce resting eggs. This process takes place everywhere in the Arctic at the same time, despite the very different temperatures prevailing in the various lakes and pools. In fact Stross (1969) even demonstrated photoperiodic induction, although the photoperiod responsible did not correspond to the day length (continuous daylight!) under which it took place. Vaartaja was able to show that birch (Betula) and pine (Pinus) grow unusually quickly if exposed to the photoperiod of their native region (Fig. 21). Krüll's discovery that fluctuations in the spectral composition of light (colour temperature) act as the diurnal zeitgeber will no doubt help to clarify the situation. Light and darkness as an "on or off" switch will have to be replaced by a new hypothesis to explain the non-linear response of the organism to equal alterations in the strength of the zeitgeber (per unit of time).

4. The Importance of the Proximity of the Ocean: Marine Birds and Mammals

Scarcely anywhere in the Arctic is the ocean far away. Regions far from the sea, like Greenland, are perpetually covered with ice. As a result of the high content of nutrients important to planktonic algae many regions of the Arctic Ocean have a high productivity and can thus support a rich fauna. Consequently marine mammals and birds are of considerable importance on arctic coasts. In addition, there is an increasing tendency with rising latitude for their colonies to move from far-lying isolated islands to inland localities (Fig. 82). On Spitsbergen some breeding colonies of the fulmar (Fulmarus glacialis) are even situated 30 km inland.

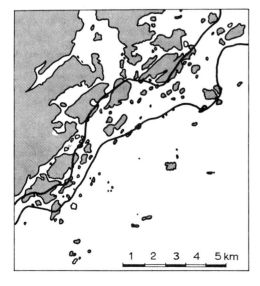

Fig. 82a. An archipelago region west of Helsinki. Only the outermost seaward zone is inhabited by marine birds. (After Bergman, 1939; adapted from Remmert, 1957)

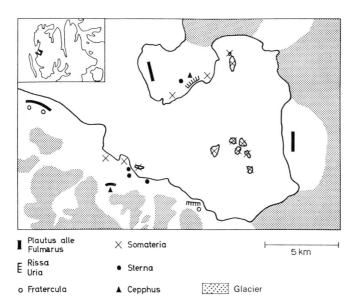

Fig. 82b. Breeding places of marine birds in Kongsfjord (Spitsbergen). (After Remmert, 1957)

The inland colonies are an excellent source of dung for the tundra not only on account of droppings but because the birds often loose food on their way back to their young. The colonies themselves show strong indications of eutrophication, especially along streams and rivers. Clearly, any discussion of the arctic tundra that does not take into consideration the significance of the proximity to the sea is unthinkable. A rapid turnover can have the same effect as eutrophication. Turnover in the tundra, however, is extremely slow on account of the low temperatures, and the soil often appears impoverished. In fact its only source of wealth often lies in the activity of the marine animals. We are indebted to Eurola for a very thorough investigation of this phenomenon on Spitsbergen (Eurola and Hakala, 1977). Studies on various bird rocks on Spitsbergen, some often far inland, revealed a very high degree of eutrophication of the tundra in their vicinity (Fig. 83), involving practically all nutrients from potassium to calcium, phosphorous and nitrogen. The eutrophication is reflected in the typical flora and accompanying characteristic fauna of the bird rocks (Hinz, 1976; Fig. 84). The data on subantarctic islands published in 1977 by Smith may be comparable and at least provide stimulation for future investigations. He reported that the most important plant society on Marion Island consists of a fern, Blechnum penna-marina. In the breeding places of Procellariidae, however, this society is replaced by luxuriant grassland dominated by tussocks of Poa cookii. The soil beneath the tussocks is much enriched in organic nitrogen and phosphorus, and the plants themselves contain more potassium, phosphorus, calcium and nitrogen than those growing outside the bird colonies. This is quite obviously due to the eutrophic influence of the birds. On the same island the resting places of seals also receive large quantities of excrement and the water draining off is rich in nutrients. As a result the vegetation in the vicinity grows well and itself has a high nutrient content (Smith,

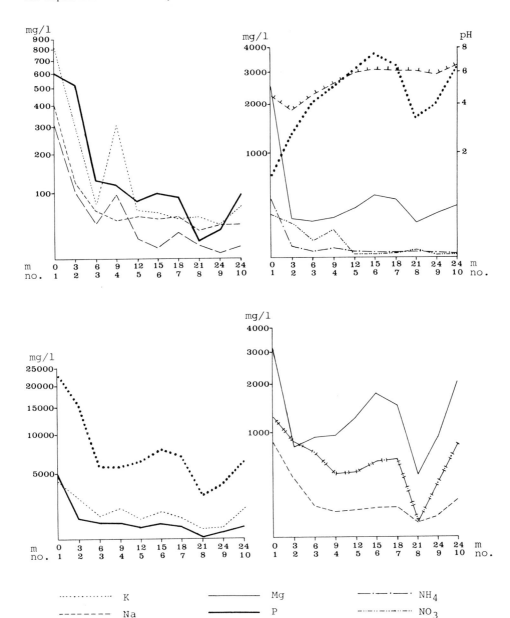

Fig. 83. Eutrophication of the tundra in the vicinity of marine bird colonies on Spitsbergen. (Eurola and Hakala, 1977)

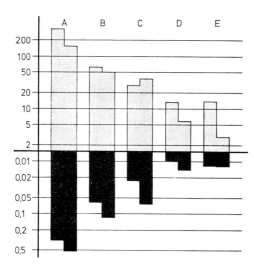

Fig. 84. Total number of individuals (*above*) and total biomass (*below*) per trap and day in Barber traps (*left*) and yellow traps (*right*) on some tundra types on Spitsbergen: *A* Diabasodden with dung from marine birds; from the Adventdalen region; *B* wet moss tundra; *C* Cassiope-Dryas-Equisetum heath in southerly position; *D* Fjell plateau with deflation heath; *E* steep slope with southerly aspect. Logarithmic presentation. (Hinz, 1976)

1977). However, Marion Island belongs to the group of subantarctic islands with the most equable climate of the world. The growth period extends practically over the entire year as a result of the perpetual oceanic subantarctic summer. We are still unable to estimate the influence of the large bird colonies in the Arctic. It is probably greater than in the Antarctic, due to the absence of flightless birds (penguins play the largest role in the Antarctic), which means that the colonies are in many cases situated farther from the water and the effect on the tundra is therefore much greater. The seals of the Arctic, the bearded seal, the harp seal and the ringed seal, spend by far the larger part of their time on the ice instead of on land. Only the walrus, which has been exterminated over large parts of its former range, and the polar bear spend much time on land, although the extent of their influence cannot at present be estimated. There is a pressing need for quantitative investigations concerning the size and positions of the breeding colonies, and the extent and intensity of the effects of dung. These remarks may provide a stimulus for future research; but a question that demands immediate attention concerns the importance of trace elements in plants. A clear indication of their significance is given by a comparison of the bird density on Iceland, with its rich basalt rocks, and that of the Scandinavian peninsula with its poor granite.

Apparently the marine birds play a larger role in north Atlantic than in north Pacific regions, although much remains to be said on this subject. The eutrophication of tundra regions by marine birds and mammals obviously depends upon their numbers, which in turn are governed by the productivity of the near-lying ocean. Hence the bird density in the north Atlantic regions is particularly high: the number and size of the bird colonies decrease from west to east along the Siberian coast as the productivity of the ocean sinks (see Fig. 152). Only in some places on the Canadian Archipelago where the ice regularly breaks up in summer is the role of birds comparable to that in the north Atlantic. The colonies are approximately as large as those on Spitsbergen: fulmar, Brünnich's guillemot and kittiwake are the most common species (Nettleship, 1977a, b; Brown et al., 1976).

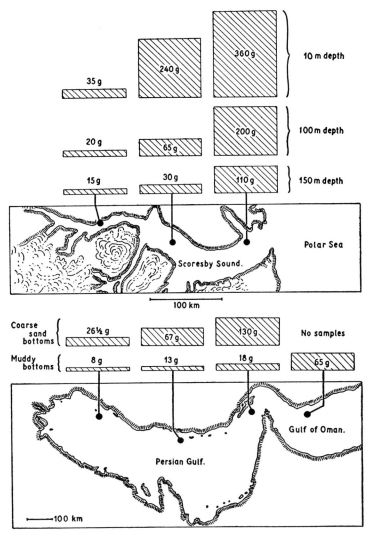

Fig. 85. Biomass of marine bottom animals in the Persian Gulf and in Scoresby Sound (Greenland). The biomass in the arctic region is far higher than that in the warm seas. (Thorson in Hedgepeth, 1957)

5. The Ratio of Productivity to Biomass in the Arctic

As already described on p. 26 and in Fig. 25 photosynthesis and catabolism differ as to their mean temperature dependence. Being at least partly a photochemical process photosynthesis has a Q_{10} of between 1 and 2, whereas that of catabolic processes is higher. This in itself means that the primary production in arctic regions must be fairly high, whereas the breakdown of the organic substance formed is a slower process. This in turn means a lower productivity of bacteria, fungi and

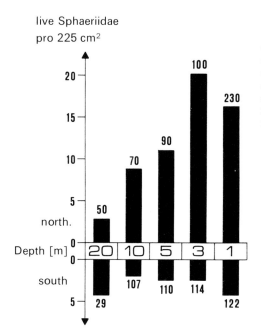

Fig. 86. Comparison of the population density of small molluscs in northern Scandinavia (*above*) and southern Norway (*below*) in relation to depth, during the summer months of 1972 to 1975. The number of bottom grab samples contributing to the total mean values appear at the end of the columns. The highest biomasses are found in northern Scandinavia. (After Hinz, 1976)

poikilothermic animals despite a rich food supply, and ought to result in a high density of ectothermic organisms because of relatively low food consumption (growth and productivity) of the individual. This theoretical prediction was confirmed many years ago for a marine milieu by Thorson in Hedgepeth, 1957; Fig. 85. The numbers of marine bottom animals in fjords in eastern Greenland are much higher than in warmer waters such as the Persian Gulf. (As everywhere, numbers decrease with depth.) This is one reason for the "rich" fishing grounds in arctic regions. Although the numbers of fish are high the fact that productivity is relatively low means that the fish population can suffer drastic reductions much more easily than in warmer waters. (In addition, the phytoplankton of polar regions is often particularly well supplied with nutrients by water rising from the depths.) Thorson's calculations showed that the overall productivity of the larger populations of the Arctic is about the same as that of the much smaller populations of warmer waters. The Antarctic is even better stocked with marine benthic organisms, probably because of the higher productivity due to better illumination of the plankton algae in the Antarctic at comparable temperatures.

This has been confirmed for other environments, too (Fig. 86), and is the only possible explanation for Hinz's finding (1975) of much larger numbers of small molluscs in freshwater lakes of northern than of southern Norway. That a corresponding situation can exist on land has recently been shown by Sendstad and Solem, who explained the unusually high density of the larvae of terrestrial chironomids in moist ground on Spitsbergen in the same way. The midges, whose larvae are barely a centimetre in length, fall into four clearly differentiated size classes, and thus require about four years for the completion of one generation.

Apparently the Arctic is a favourable environment for all warm-blooded organisms. The adaptive advantage of endothermy as opposed to ectothermy becomes even more obvious than in warmer regions. This includes the digestibility of plant matter (see p. 33) as well as the ability to exploit the large numbers of slow-moving ectothermic organisms. The striking dominance of marine birds (guillemots, penguins, fulmars) and marine mammals (seals, toothed and bearded whales) in polar regions can be explained convincingly in this way, as well as the paradoxical fact that large species like the whale and seal can control the numbers of very small animals like krill and fish (Remmert, 1978).

This is the reason for the abundance of trout and salmon in arctic rivers, and for the frequent impoverishment of the waters in holiday areas: once the rivers have been overfished recovery takes many years.

The erroneous assumption that high biomass is synonymous with high productivity has led to some dangerously false conclusions. In actual fact the productivity of tropical seas is probably not less than that of arctic waters, although for the reasons already mentioned the biomass of the former is considerably smaller. A complete revision of our views in connection with productivity research will probably be necessary (El-Sayed and Turner, 1977; Knox and Lowry in Dunbar, 1977).

6. Species Problems

The limits of distribution and the classification principles commonly employed by zoogeographers are largely inapplicable in the Arctic as a circumpolar region. The majority of genuinely arctic species are of circumpolar distribution, a situation described by the zoogeographical term "holarctic" and connected with certain problems. For example, if very different subspecies meet at the extreme ends of their range of distribution, they may behave as individual species. This phenomenon is encountered in many other contexts, and is no longer considered to be a great problem. Of far more importance is the fate of arctic organisms in the glacial period. The statement that they were pushed further south during the pleistocene and later reestablished themselves where we find them to day is an over-simplification of the situation. The distance to which the ice-cap extended southwards from the pole in the Northern Hemisphere was by no means everywhere the same (Figs. 87, 88). Broad expanses in north Siberia, all of north Greenland and large areas around the Bering Sea, which are today covered by what we term arctic tundra, were also polar tundra when the ice-cap was at its maximum. Obviously such ice-free remnants of tundra retained their own fauna, so that in fact the arctic fauna was not simply displaced to the south during the pleistocene but was split up and isolated in a large number of ecologically different areas. It was these faunal remnants that recolonized the present-day polar regions after the glacial period. Populations from very different regions now met up again (Fig. 89). In a number of cases involving European birds it can safely be assumed that the two forms, as for example the European nightingale and thrush-nightingale, or the carrion crow and hooded

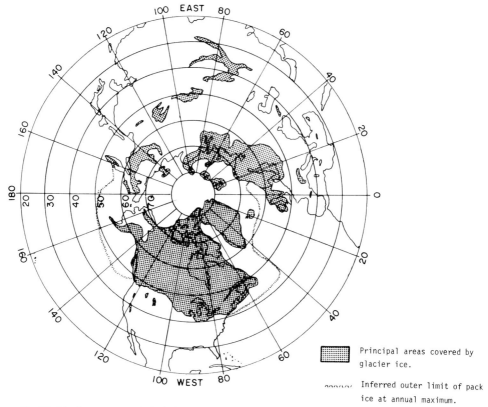

Fig. 87. Maximum extent of glaciation and pack ice during the glacial period. (After Flint, 1957 from Irving, 1972)

crow, developed in different refugial areas although they have not merged again since. In both of these cases the present-day ranges of distribution of the species pairs are relatively well separated but those of the tree-creeper and the short-toed tree creeper, a pair that must have evolved in a similar manner, are partly identical. It would be expected that a splintering into widely differing regions would have led to very marked species formation. However, time was apparently insufficient and what happened more frequently was that forms that had been on the point of speciation again fused together. As a result the systematics of these animals are totally confusing, a classical example being provided by the reindeer. American texts generally distinguish between four species: the Greenland reindeer, relatively similar to the European form; the Peary reindeer, a rather short-legged animal living in fairly small herds; the barren-ground caribou, which forms the familiar vast herds; and the forest caribou that usually lives on its own (Fig. 91). As Fig. 90 shows, the subdivisions could be carried much further. The four species came largely from different refugia, some of which are depicted in Fig. 92b. We know far less about such refugia and possible routes of recolonization in the Old World. Here too, however, a forest reindeer is everywhere encountered. It has never been

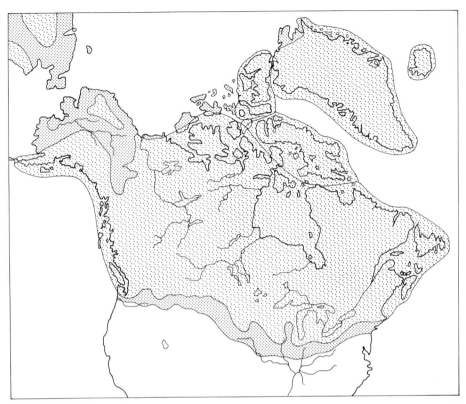

Fig. 88. Approximate distribution of tundra (*heavy stippling* excluding areas now submerged) and continental ice sheets at the Wisconsin glacial maximum in North America. (MacPherson, 1965)

domesticated and lives alone or in small groups. The Fjell reindeer, the only form that has been domesticated in the Old World, often penetrates into the forests in winter (like the barren-ground caribou) and lives alone or in small groups like the forest reindeer. The two forms reputedly cannot be crossed, although it is difficult to find clear-cut and consistent differences between them. Although the extremes of the two forms can be well distinguished there are just as many transitional forms. The complexity of the situation is illustrated by the following argument. If forest caribou and forest reindeer are in fact distinct species they must have evolved in different places and their similarities are thus analogies. Each of them must therefore be an individual species, just as the Fjell reindeer is. Their similar behaviour, appearance and physiology are merely the result of living in similar environments. If we assume that the present-day range of the North American caribou was colonized from 3–5 refugia, a far larger number of refugia would have to be postulated for Eurasia. Although this concept explains the complications involved in the systematics of the reindeer it does nothing towards disentangling them. From what has been said it becomes obvious why the systematics of so many arctic animals has been interpreted in such very different ways. The gyr falcon, for example, is treated in most recent publications (Glutz) as a polymorphic complex,

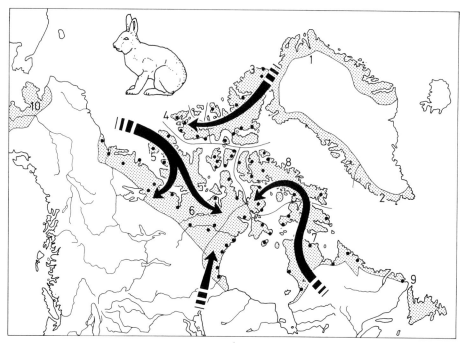

Fig. 89. The recolonization of the tundra of North America after the ice had receded; based upon present-day views concerning races and a knowledge of the species inhabiting regions that remained free of ice during the glacial period. (MacPherson, 1965). **a** Arctic hares (Lepus): *1* L. arcticus groenlandicus; *2* L. a. porsildi; *3* L. a. monstrabilis; *4* L. a. hubbardi; *5* L. a. banksicola; *6* L. a. andersoni; *7* L. a. labradorius; *8* L. a. arcticus; *9* L. a. bangsi; *10* L. othus (= timidus)

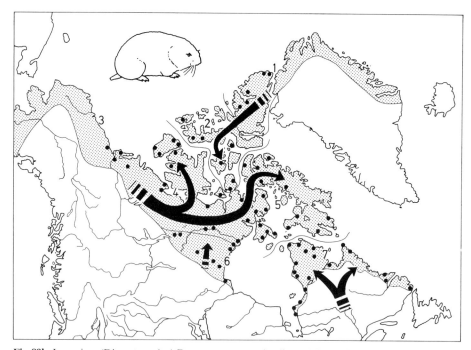

Fig. 89b. Lemmings (Dicrostonyx): *1* D. torquatus groenlandicus; *2* D. t. clarus; *3* D. t. rubricatus; *4* D. t. lilangmiutak; *5* D. t. lentus; *6* D. t. richardsoni; *7* D. hudsonius

Species Problems

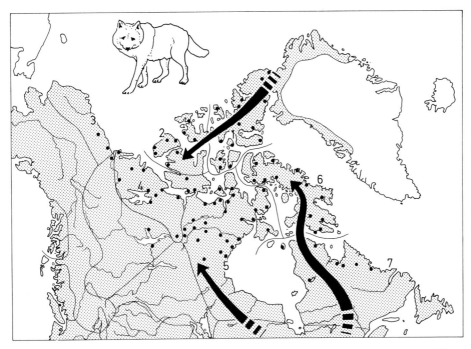

Fig. 89c. Wolves: *1* C. l. orion; *2* C. l. arctos; *3* C. l. tundrarum; *4* C. l. maceril; *5* C. l. hudsonicus; *6* C. l. manningi; *7* C. l. labradorius

although it was previously subdivided into a series of very different subspecies. The same applies to the different colour forms among the guillemots, fulmars and skuas (Stercorarius).

The large European gulls provide a good illustration of just how complicated the systematics of arctic animals can be. The generally accepted view is that subsequent to the glacial period there were two separate populations of related herring gulls: the yellow-footed form in the Asiatic inland regions, and the pink-footed in the inland and coastal regions of North America. The latter spread throughout North America and across the Atlantic to Europe and were the ancestors of our familiar herring gulls (Larus argentatus), of which there are two races, argenteus in Ireland, the British Isles, western Norway, France, Spain, and the north German Bight, and argentatus in the Baltic. Coming from the east, a group of yellow-footed forms from the interior of Asia made its way into the same region and pushed forward into the Mediterranean (michahellis), the Atlantic islands (atlanticus), the North Sea (including the Norwegian coast, the north Atlantic islands and Iceland: (britanicus-graellsii) and Denmark (intermedius) as far as the Baltic (fuscus). The three forms fuscus, intermedius and britanicus have a relatively dark back and are known collectively as lesser black-backed gulls. Atlanticus is also sometimes included in this term. The use of "black-backed gull" as if this were a single species is probably due to the fact that, although the herring gull and the lesser black-backed gull often live side by side, bastards have hardly

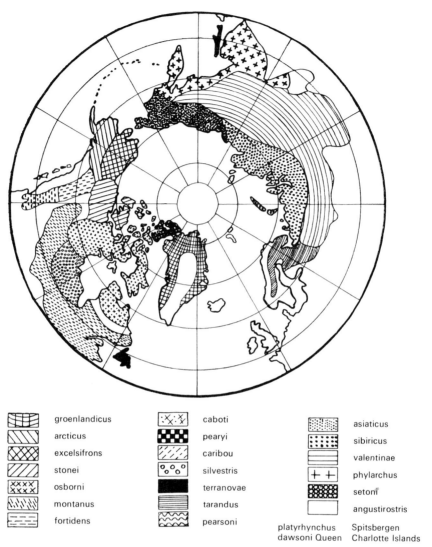

Fig. 90. Sub-species of reindeer and their geographical distribution. (After Herre, 1956.) There are no native terrestrial mammals on Iceland except polar fox

ever been observed. The two are also ecologically well separated, the black-backed gulls mainly inhabiting rocky coasts and small, rocky offshore islands, whereas the herring gulls mostly prefer the dunes. The eastern black-backed gull (fuscus) is a strict migrant, taking inland routes as far as Lake Victoria; the western forms (britanicus and intermedius) follow the coasts as far as equatorial Africa, whereas the herring gulls remain for the most part in Europe.

This situation has been described many times, although only inadequately. The forms that advanced westwards through Asia also spread out along the coasts of the Arctic Ocean, on the White Sea, and along the rivers, marshes and lakes south of the

Species Problems   115

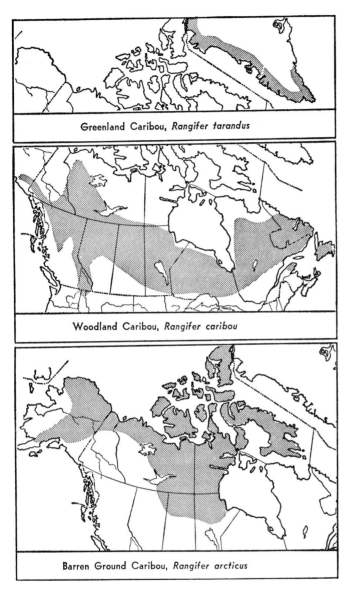

Fig. 91. The distribution of reindeer "species" in North America. (After Burt and Grossenheider, 1964)

Gulf of Finland (Fig. 93). These forms are lighter in colour, more like the herring gull, but on account of their yellow feet they have to be grouped with the black-backed gulls. The subspecies occurring in European regions are variously named. They are often lumped together as cachinnans, and for the sake of simplicity this is what we shall do here. About 100 years ago cachinnans advanced as far as Finland, with the result that in the region of the Gulfs of Finland and Bothnia three large forms of gulls existed side by side: cachinnans in marshes and marsh lakes,

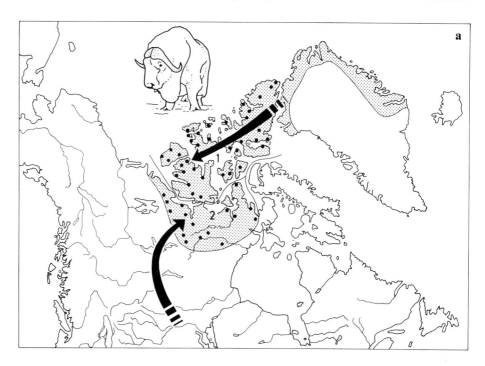
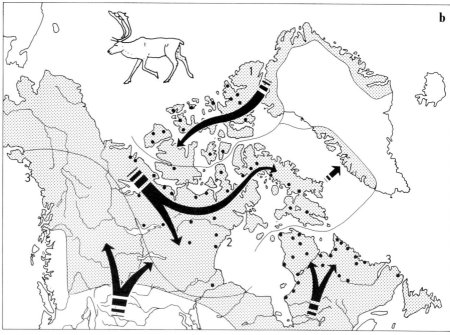

Fig. 92a,b. As Fig. 89. **a** Muskoxen: *1* O. m. wardi; *2* O. m. moschatus. **b** Reindeer: *1* R. tarandus ? pearyi; *2* R. t. groenlandicus; *3* R. t. ? caribou. *Broken lines* indicate intergrade population on western arctic islands, and seasonal overlap on western mainland

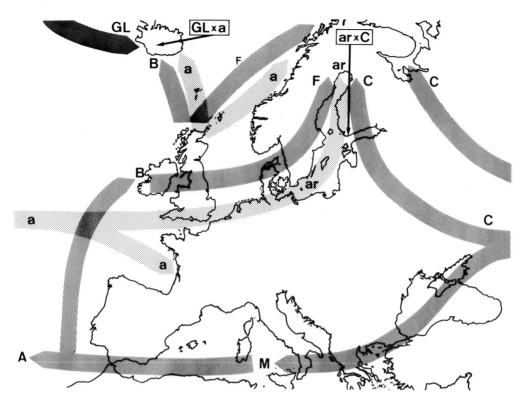

Fig. 93. Related forms of herring gull and lesser black-backed gull breeding in Europe, and the paths taken in the course of their distribution. The occurrence of breeding glaucous gulls (Larus hyperboreus) as well as the regions in which bastardisation occurs are also indicated. *Hatched, capital letters*, yellow-footed forms: *C* cacchinans; *B* brittanicus; *M* michahellis; *I* intermedius; *A* atlanticus; *F* fuscus. *Dotted, small letters* pink-footed forms: *a* argenteus; *ar* argentatus. *Cross-hatched, two capital letters: GL* glaucous gull, L. hyperboreus. (Remmert, 1980)

argentatus along the coasts, chiefly in dune areas, and fuscus also on the coasts but mainly on the rocky islands. Small numbers of argentatus were also seen on the large Finnish lakes in places suitable for breeding. Hartert had already drawn attention to this latter observation and it has been mentioned repeatedly by Voipio in recent papers.

As everywhere in the wake of Man, the number of herring gulls also increased in the Gulfs of Bothnia and Finland. Fuscus has even been ousted by argentatus to some extent, the latter spreading in ever-increasing numbers on the inland sea, and cachinnans on the coast. The Finnish herring gull population today is a mixture of argentatus and cachinnans. The process began about 40 years ago and is practically complete. No preferences whatever are shown in mating and a perfect state of panmixy exists between the two, despite the fact that cachinnans are more closely related systematically to fuscus than to argentatus. Mixing has taken place here between the two forms most closely resembling each other externally and ecologically, despite the fact that they are not the most closely related forms. Their ecological similarity is based upon the fact that both cachinnans and argentatus find their food to a large extent on land and in coastal regions and thus have a much closer affinity for civilization than fuscus, which seeks its food chiefly on the open sea. The fusion of two differing populations takes place whilst a third population behaves as an individual species, although phylogenetically very closely related to one of the two merged populations.

In the course of its enormous increase in numbers the herring gull (argenteus) has also invaded Iceland from the south-east, concurrently with approximately equal numbers of britanicus. The two species are ecologically well separated. Whereas argenteus prefers the vicinity of human beings, britanicus seeks its food at sea. Two large species of gull, the great black-backed gull, Larus marinus, and the glaucous gull, Larus hyperboreus (= glaucus), were already established in Iceland. Both are very much larger than the lesser black-backed gull and the herring gull, and have sometimes been jointly allocated a subgenus of their own. Marinus is usually an asocial bird and is not attracted to civilization, preferring to seek its food at sea or on bird rocks. Hyperboreus, however, has attached itself to Man throughout the Arctic and like the herring gull in central Europe it frequents rubbish dumps. Hyperboreus and the much smaller newcomer argentatus thus have very similar ecological requirements. Ingolfson showed that nowadays neither hyperboreus nor argentatus are found in Iceland: they have been replaced by a bastard population of the two species. Only in the extreme north-west of Iceland, not far from Greenland, are there probably still pure hyperboreus colonies. This is a further example of the merging of different populations. In this case very different species (even considered at one point to belong to different subgenera), faced with almost identical environmental requirements, have formed bastard populations.

The existence of such very different populations of arctic animals is probably one of the reasons for the very varied reports concerning their nutrition. Whereas data on the winter food of reindeer in Europe and Asia or of the caribou on the North America continent and the islands of Hudson Bay invariably indicate the predominant role of lichens (for America e.g., Parker, 1972, 1975; Miller, 1974, 1976; Miller and Broughton, 1974; Dauphine, 1976), these appear to be of no importance whatever to the reindeer of the Canadian Archipelago. The distinctly stationary populations of polar bears in the southern part of Hudson Bay appear to be pure herbivores in summer, living off vascular plants, marine algae and berries, whereas the populations on the islands of the Canadian Archipelago prefer small mammals, birds and eggs. The High-Arctic populations, and those of Spitsbergen in winter, depend on seals as a source of food (Manning, 1971; Harington, 1973; Yonkel et al., 1976; Stirling et al., 1977).

Similar population situations can be found among fish, and particularly interesting examples are the arctic genus Salvelinus and the subarctic Coregonus. It can be safely claimed that every lake has its individual and more or less genetically fixed form. However, conditions are more complicated than for mammals and birds, and apparently no attempt has ever been made to reconstruct the routes taken by fish in post-glacial times. Investigations on insects are even less satisfactory and should be regarded with the utmost caution or even scepticism. In some cases, species names are of relatively little value for ecological purposes unless specimens are available for inspection. In the case of marine organisms, on the other hand, the type of distribution is quite different, North Atlantic and North Pacific being almost completely separated. This applies equally to fish (Salmo in the Atlantic, Oncorhynchus in the Pacific), to seals (sea-lions only in the Pacific) and to birds, which apparently indicates the great length of time for which the two systems have been separated. The birds have attracted particular attention: some investigations have dealt with niche occupation, involving theoretical ecological questions. Of the six species of Alcidae that breed in the North Atlantic and North Pacific only one is common to both (Uria aalge, the guillemot). Of the remaining five some are sibling species and some are only distantly related to one another. Nevertheless, there is a high proportion of parallel niches, as was stressed by Cody (1974). (Some of his findings have to be re-evaluated: the little auk, for instance, seeks food in the neighbourhood of its nesting area, with the cliffs always in sight, and only after the breeding season do the birds venture far out to sea.)

In connection with migratory birds that travel as far as the Antarctic the question arises as to whether some individuals also breed in their winter quarters and establish a population. In fact it appears that a mechanism of this kind was involved in the formation of species with bipolar distribution, although whether the species spread from the north or from the south cannot be discussed at this point. Good examples are provided by the fulmar, with a well-defined subspecies in the Antarctic, and the great skua (Catharacta skua), with several well-defined subspecies on Antarctica and the offshore islands. Terns apparently also fall within this group since the antarctic form, Sterna vittata, is very similar to Sterna paradisea, the arctic tern. Increasing attention is being paid to such similarities at present. The southern fulmar has at times been placed in a genus on its own, although the probable explanation is that the investigators concerned were specialized to a particular geographical region and were less familiar with related species on other continents. Northern and southern fulmars are in fact not easily distinguished. Surprising observations have been reported on great skuas. Individuals that had been ringed in the Antarctic whilst breeding have been caught in the Northern Hemisphere, and showed little or no differences from individuals breeding in the north. At present it is not known whether there is a gene flow between the northern and southern populations. Sterna paradisea and S. vittata, however, appear to be quite clearly separated.

A problem so far untouched is presented by the many populations of primary marine animals living in the freshwater lakes of the Arctic. Following the melting of the ice at the end of the glacial period, the land almost everywhere became higher, and what had previously been salt-water bays slowly turned into freshwater lakes in many of which marine animals can still be found today. The most interesting are Mesidothea entomon, Pontoporeia affinis and Mysis oculata relicta. Secondary marine animals like the ringed seal (in Lakes Baykal, Kaspi and Saima), the harbour seal (in Labrador) and the fish Cottus quadricornis, have all adapted to the osmotic and static conditions of freshwater, which at least the ancestral forms of the primary marine animals could not. The wealth of populations that thus came into being were sexually completely isolated from the ancestral species and from one another. According to the current definition of a species (an area of gene diffusion), each population would have to be regarded as an individual species, which of course does not make sense. What we are confronted with is identical adaptation (osmoregulation), or multiple identical species formation over a very extensive area. It would be interesting to investigate whether the evolution of these species is taking a different course in the different lakes as a result of local biotic and abiotic condidions.

In general, the diversity of flora and fauna decreases from south to north in the Northern Hemisphere. The number of species of all insect groups, plant groups, fish, amphibia, reptiles, birds and mammals is higher in the temperate regions of America and Eurasia than in the Arctic regions. Historical reasons are usually put forward to explain this phenomenon, e.g., that insufficient time has elapsed since the diappearance of the ice for a highly diverse flora and fauna to develop. This is no doubt true of some places, particularly of isolated islands such as Iceland and Spitsbergen, although there are two arguments against this view:

1. Marine warm-blooded animals have achieved an astounding species diversity. This is seen in seals, whales and sea-birds and applies equally to the Antarctic and Arctic. The narrowness of their niches calls to mind the African ungulates, also characterized by their very narrow niches.
2. Not all polar regions had to be recreated after the Ice Ages. On both hemispheres certain regions remained free of ice and have been in continuous existence for a very long time. Spatial and temporal continuity are of much longer standing than is generally conceded in discussing the biological age of northern polar regions.

Therefore a physiological explanation of the low diversity of arctic flora and fauna seems to be the most plausible. This would mean that this region represents the physiological limit for most forms of life and only a few species can exist at all. Economists have shown that at low (human) population densities only a few specialized forms of occupation are possible; at high densities specialization (niche formation) gradually increases until a wealth of highly diverse occupations is established to satisfy the demands of the population. An analogous situation would exist at the physiological limits of life. Only warm-blooded animals have not reached their physiological limits in the Arctic and this is the only property that would render a certain amount of diversity possible. Diversification would be most likely in regions where food is available all year round, for example on and near the oceans. In fact a remarkable degree of diversity is encountered in arctic and antarctic marine birds and mammals, in very narrow, widely overlapping niches. A physiological explanation of species formation is in substantial agreement with Enright's hypothesis that climatic factors, which have so far been regarded as not affecting density, do in fact exert a very decisive influence on the population density of a species (Enright, 1976).

Many attempts have been made to introduce arctic animals into other regions. Arctic hares have unsuccessfully been set loose on Spitsbergen on a number of occasions, and muskoxen have (successfully) been introduced into southern Norway and on to Spitsbergen (extinct in most recent years), but attempts in Iceland, northern Norway, Scotland and many other places have failed. Much attention has been attracted by the experimental introduction of the Scandinavian domestic reindeer into Iceland (Thorsteinson et al., 1970), onto islands in the Bering Sea (Scheffer, 1951) and on South Georgia (Kightley and Lewis Smith, 1976;

Leader-Williams, 1978). In each case it was assumed that the animals would reproduce rapidly and provide the inhabitants with an additional source of food. Reindeer were introduced into three localities in Iceland as early as the eighteenth century. Of these, only one herd in the eastern central highlands has survived. The other regions are probably too oceanic for the animals. Today, the herd numbers about 3000 individuals and seems at no time to have been particularly large. Detailed studies concerning its influence on the environment have not been carried out. The experimental introduction of reindeer onto islands in the Bering Sea ended in a familiar catastrophe. The animals multiplied rapidly at first and over-exploited the islands, after which the populations very suddenly collapsed to zero or thereabouts. The reindeer transplanted to South Georgia had already changed their reproductive period in the second year after their arrival to correspond to the reversal of the seasons. On the whole, reproduction was slower than in other places, which may have been due to more intensive hunting as compared with Iceland and the Bering Sea islands. The native plants (Poa flabellata, Acaena magellanica) appear to be highly sensitive to grazing. The high tussocks of Poa flabellata reach up to the surface of the snow and are thus readily available to the animals, so that a large part of the buds for the following spring is destroyed. For this reason an imported species, Poa annua, is increasingly taking over in intensively grazed areas.

IV. Peculiarities of the Systems

1. General Principles

The transmission of energy and matter is of primary importance in modern ecosystem analysis. In order to determine these two functions in animal populations a knowledge of a variety of parameters is essential. These are:

1. The number of animals per unit area during the period under investigation, and their mortality.
2. The growth of the animals, i.e., their production of organic substance in the period investigated.
3. Their food consumption and the passage of this food, i.e., whether it is respired or immediately used for the production of organic substance, whether it is digested or expelled again as faeces. Above all, it is the ecological efficiency of the organisms that is of interest, in other words, the ratios of consumption to production, of consumption to assimilation and of assimilation to respiration (see Remmert, 1978).

Let us first of all consider points 2 and 3. Obviously the values differ from one species of animal to the next and must be determined separately for each. Although at first sight such values would appear to be easily determined they are almost unobtainable in practice. A number of reasons are responsible for this. In the first place, a large element of uncertainty is involved in the application of analyses of food requirements under laboratory conditions to the situation in nature. Laboratory studies seldom cover a longer period of time, besides which it is almost impossible to employ the temperatures prevailing in the open, and ecological efficiency is highly dependent upon temperature. There are additional complications due to the cooling effect of the wind, which causes a rise in energy consumption and a drop in productivity. Growth of poikilothermic animals in the Arctic is extremely slow (Table 2), and uninterrupted observation is impracticable. In experiments involving warm-blooded herbivores it is impossible to offer the specific food items available in the field. Differences in digestibility may also lead to erroneous results. Furthermore, food requirements, ecological efficiency, growth and productivity exhibit a characteristic annual pattern in animals with a longer life span. All take up less food in winter (even if available) than in summer, and its digestibility also fluctuates over the course of the year. In a few rare and exceptional cases, for example the snowy owl, it has been possible to arrive at an exact estimate of food requirements in the field and to compare them with laboratory values (Table 5). This brings up the question of whether the food requirements of the snowy owl in the open are the same as its consumption. The values were obtained at a lemming peak, when the birds expend very little energy in catching their prey. In such years many more individuals are killed than are actually consumed (Ceska, 1978), and these are not included in Watson's calculation. Thus, on the basis of such observations the owls' ecological effect ought to be trebled. In the case of herbivores the situation is similar. Their food is difficult to break down, so that although they may take up very large quantities only a small proportion is absorbed. The faeces production of such animals is very high, but the organic substance produced is very little considering the amount of food consumed. On the other hand, the minerals are washed out of their faeces much more rapidly than from living plant matter. Primary and secondary decomposers in the soil can attack and remineralize litter containing faeces more readily and effectively than simple plant matter that has not previously been masticated. The large quantities of faeces therefore speed up the nutrient cycle within an ecosystem, although this is not immediately apparent. It is reported that the reindeer in southern Norway destroy about three times more vegetation by trampling than they consume.

At present we can merely speculate about the ecological efficiency of fish under High-Arctic conditions. In general, it seems that predatory fish can expend 20%–40% of the energy consumed as food on production, which is an extremely high value as compared with other animal groups. These figures were obtained from a large number of laboratory investigations as well as from complicated field studies in both salt and freshwater, with subsequent computer calculations. According to Arntz (1974), in cod from the Baltic about 60% of the energy intake is expelled in an unused form or respired, and 40% is available for production. In herbivorous fish, however, only 2% of the energy intake is used for production. This low value is due to the difficulty of assimilating plant matter as compared with animal food. Since the ease with which plant matter is broken down is greatly influenced by temperature, herbivorous fish can be excluded from High-Arctic regions. It can therefore be assumed that the ecological efficiency of the fish of the High Arctic is about 40%, but there is a pressing need for more exact data an this point.

The overall picture is roughly as follows: predatory poikilotherms assimilate about 80% of their food intake. About half of this is available for growth (=production), and the rest is respired. This is the situation in marine worms (Nereis), spiders, predatory insects and predatory fish. Poikilothermic herbivores, on the other hand, assimilate at most 20%–40% of their food intake, of which, again about half is available for production. However, assimilation can be considerably lower, in which case production is also lowered in relation to consumption. Examples are provided by herbivorous insects, animals living in the ground litter and herbivorous fish. Homoiotherms, whether predatory or herbivorous, generally assimilate 80% of their food, and only in certain herbivores whose food is very difficult to digest is the figure as low as 30%. Nevertheless, due to very large respiratory losses, only 0.5%–2% of the energy intake is available for production.

Young animals, often constituting the larger part of a population, are of special importance in ecosystem analysis (Fig. 94). Studies carried out by Pöhlmann on domestic geese and by Ceska on young snowy owls both showed that the young consume considerably more food per gram of body weight than the adults. However, this relationship is the same as that existing between species of different sizes. In fact the growth of the young birds appears to depend on their better utilization

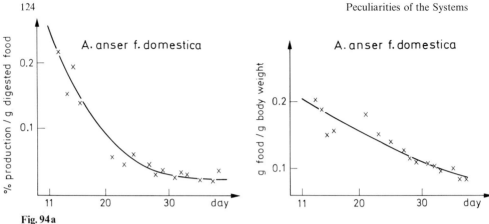

Fig. 94a

Fig. 94a, b. Food intake of young and adult animals. **a** Production per gram of digested food in grey-lag geese (left), in percent of body weight. Food intake of young, growing grey-lag geese related to body weight (right). (Pöhlmann, 1976). **b** Daily food quantities, in percent of body weight, in growing snowy owls and other species of owl of the same size (*top*); food intake per day in grams dry weight with increasing growth in snowy owls (*centre*); and weight increase in young snowy owls in the Nürnberg Zoo (*bottom*). (Ceska, 1978)

of the food consumed. In young snowy owls food is utilized to 94% during the first days of life and to 85% at about 14 days whilst in the adults the figure drops to 55% (Ceska). The ecological efficiency of young animals is thus not the same as that of adults of the same species, as Steigen, for example, also found in the case of spiders (Steigen in Wielgolaski, 1955; see also Phillipson, 1966). The same relationship applies throughout the animal kingdom. Small specimens of the mollusc Arctica islandica utilize their food to 75%, whereas in adult individuals the value drops to 43% (Winter, 1969).

Some idea of energy requirements can be obtained using the familiar relationship between metabolic rate and size (Fig. 95), from which it follows that the available food in any environment is much more rapidly consumed by small organisms than by the same biomass of larger animals. Nevertheless, the situation is already more complicated in the case of marine birds and mammals. If the basic metabolism is calculated per kg body weight and the results compared with the quantity of food consumed by zoo specimens (also calculated per kg body weight), the discrepancy between the two values increases with decreasing body size (Fig. 96; Table 7). This implies that in the case of homoiothermic marine vertebrates the formulae commonly used for basic metabolism are not valid and the index ought probably to be modified. The original simple assumption that the basic metabolism of aquatic mammals is twice that of terrestrial mammals is not supported by feeding data obtained on zoo animals. It seems that despite the considerable activity developed by some of these aquatic mammals in captivity they still do not eat sufficient to achieve twice the metabolic turnover of terrestrial forms.

Therefore the question arises as to how to determine the quantity of energy required under the more rigorous conditions prevailing in natural surroundings from the calculated values for basic metabolism. The figures for terrestrial animals vary, according to species, by a factor of from 3–4 for the roe deer, Capreolus capreolus, to 1.1 for spiders, Lycosidae. In feeding experiments the ease or difficulty

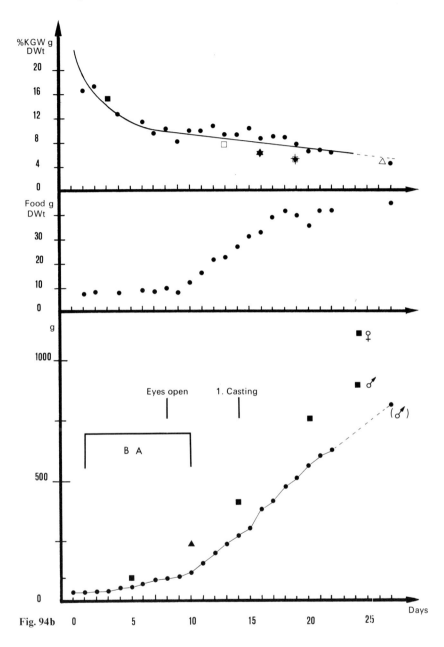

Fig. 94b

with which the food can be broken down has to be borne in mind. In predatory forms the ratio of consumption to assimilation (between 80% and 90%) is probably always the same, whereas in poikilothermic organisms the ratio of consumption to assimilation is very unfavourable at low arctic temperatures. However, it should be borne in mind that arctic flies have a higher fat content and thus a higher caloric content and nutritional value than those of temperate latitudes, so that a simple

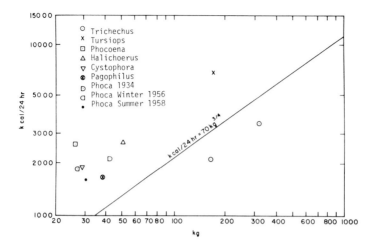

Fig. 95. Resting metabolism of marine mammals. It is about twice the value given by the formula generally employed, and only in the case of the very sluggish seacow, that inhabits warm waters, is it lower. (Irving, 1972)

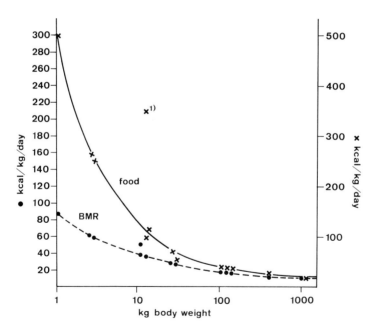

Fig. 96. Daily food intake of marine birds and mammals in zoological gardens (food requirements in calories per gram body weight), compared with the calculated basal metabolic rate (BMR) of these species. (^{1}) This very high value applies to the king penguin kept continuously at temperatures around freezing point in the Frankfurt Zoo

estimate based on dry weight can give misleading results. The warm-blooded animals of the Arctic may also exhibit a very unfavourable ratio of consumption to assimilation on account of the indigestible nature of their food. From the quantity of faeces produced per unit of time by a warm-blooded herbivore in an experiment, and the quantity produced (by a known number of individuals) per unit of time in the open, it is possible to deduce the digestibility of the food consumed under the latter conditions. For example, a reindeer in a zoo produces about 100 g of faeces daily on concentrated feed, whereas on Spitsbergen 670 g are produced, with a combustion value of 3,612 kcal per g). A 55 kg reindeer consuming 6,400 kcal concentrated feed daily in the Nürnberg Zoo would require at least 11,100 kcal of natural food per day on Spitsbergen. Due to the very low nitrogen content of the plants a still larger consumption has to be assumed. A very similar situation is found in poikilothermic animals of the litter of temperate forests. They need extraordinarily large amounts of food but assimilate only a very small part of it. Thus mycetophilids and sciarids chew and consume about 30% of the beech leaves falling in one year although they only assimilate 1%. The role of such organisms cannot be estimated too highly since they bring leaves into the only form in which they can be attacked by bacteria and fungi. This is probably the main function of arctic soil organisms.

Certain difficulties arise in connection with point 1 with regard to population analysis. In any analysis of the role animals in handing on material and energy in the system an important criterion is the size of the population. (That this criterion alone is insufficient has already been emphasized.) For piokilothermic organisms data on growth, duration of life, time required to achieve sexual maturity, and the number of eggs produced are also essential. In the case of warm-blooded forms such as birds, for example, it is necessary to know the number of territorial and non-territorial individuals, the number of eggs actually hatched in relation to the number produced, the number of young reared and their mortality. Since extensive data of this kind are not yet available for any single ecosystem, we have to make use of the size of the population. Some idea of the population density of various animal groups in arctic terrestrial environments can be obtained from Tables 8, 16, 17, 19, 20. Unfortunately the tables are highly unsatisfactory and give only the roughest impression of population densities. One of the reasons for this is the extraordinarily large fluctuations in the populations involved, which is why mammals were not included and why it is doubtful whether birds ought to have been included at all. Passerine populations, for example, seem to oscillate synchronously with the small rodents, although on a much smaller scale, (Fig. 63), with a maximum as a rule only three times as high as the minimum. Another difficulty is presented by the wide variety of measuring methods employed, and in some cases it is even uncertain whether the author has observed the generally accepted international convention of territories (=pairs) per unit area, or whether he is referring to individuals. Criticism with respect to methods applies even more with regard to poikilothermic soil animals, where extraction from the soil has been performed in a variety of ways. Each method has its errors and since the population density of soil animals within a single environment is subject to large fluctuations from one experimental area to another, the results indicate little more than that mites, enchytraeids, Collembola and Diptera larvae are by far the most numerous macroscopic animal groups in the

Table 8. Population density of birds in various arctic regions. Further data (Canada) see Erskine (1971, 72, 76, 1980)

Locality	Territories (km²)	References
NWT Canada, Tundra, 63° N	39	Höhn (1968)
Devon Island, Canada, 75° N	8–11	Bliss (1977)
Greenland, Scoresby Sound, 71° N	1.3	Hall (1966)
Iceland, central oasis 65° N	100–300 [a]	Gardasson (1975)
Hardangervidda, Norway, above tree line, 64° N	25–90	Lien et al. in Wielgolaski (1975)
Pine forest (poor), Sweden Stora Sjöfellet 67° N	48	Andersson et al. (1967)
Pine with deciduous trees Stora Sjöfellet	129	Andersson et al. (1967)
Subalpine birch forest Stora Sjöfellet	295	Andersson et al. (1967)
Birch forest, Kevo, Finland 70° N	128	Haukioja et al. in Wielgolaski (1975)
Rich spruce, Amarnäs, Sweden, 66° N	171	Enemar et al. (1965)
Amarnäs, Sweden, 66° N, heath birch	166–175	Enemar et al. (1965)
Amarnäs, Sweden, 66° N, mixed wood	217–318	Enemar et al. (1965)
Amarnäs, Sweden, 66° N, meadow birch	275–349	Enemar et al. (1965)
Moorland with pine 66° N, Finland	62–66	Väisänen and Järvinen (1977)
Spruce forest, N. Finland	107	S. Sulkava pers. communication
Moist pine forest, N. Finland	41	S. Sulkava pers. communication
Dry, poor pine forest, N. Finland	11	S. Sulkava pers. communication
Moorland with pine, N. Finland	52	S. Sulkava pers. communication
Wet moors without trees, N. Finland	70	S. Sulkava pers. communication
Wet moors with ponds, N. Finland	40	S. Sulkava pers. communication

[a] Far more birds breed here. The figures indicate only the territories (= pairs) of the pink-footed goose, which is the most numerous species

soil (only exceeded by nematodes and unicellular organisms). Data obtained with the photoelektor give a relatively reliable picture and the best agreement.

In the northernmost arctic tundra regions, i.e., Greenland, Devon Island and Spitsbergen, bird population densities are very low (except for the colonies of marine birds on the coasts). Further south the number of pairs encountered per km² increases substantially although the extraordinarily high densities found in Iceland are an exception. In Thjorsarver on central Iceland 100–300 pairs of the pink-footed goose may breed per km²; the average number of young reared is close to two. In addition species such as snow bunting, meadow pipit, whooper swan, long-tailed duck, harlequin, red-throated diver, whimbrel, red-necked phalarope, purple sandpiper, ringed plover, arctic tern, great black-backed gull, arctic skua and ptarmigan are less numerous but by no means rare. Two other species, the dunlin and the golden plover, are almost as numerous as the pink-footed goose. According to Gardasson a total of between 200 and 600 nests can therefore be expected per square kilometre. Figures of this kind indicate densities approaching those in rich forest regions. In general, bird densities in the arctic tundra are only slightly less than those recorded from poorer pine forests with very little undergrowth. Richer pine or spruce forests can support much greater numbers, but the localities with the highest bird densities of all are the wet birch forests of northern Scandinavia.

Clearly, any attempt at a general estimate of energy flow based on animal density is particularly difficult if, as is frequently the case with wild geese, the animals breed in densely concentrated colonies. In Iceland, for example, this applies to the pink-footed goose, and is even more marked in the case of the snow goose (Anser caerulescens) in the Canadian Arctic. Colonies of the latter are known to contain about 170,000 nesting pairs and about 90,000 non-breeding individuals (in the Hudson Bay region for example). A thorough investigation of the role of such colonies in the ecosystem of the arctic tundra has apparently not yet been undertaken.

Throughout the High Arctic very severe fluctuations in numbers of the populations of caribou and muskox occur in connection with extreme years, which can only be survived by a large population (Tener, 1972; Miller et al., 1977). Many of the reports available, however, contain contradictory data, which is not surprising in view of the size of the area involved. In contrast to earlier data, which indicated very large fluctuations in the numbers of barren-ground caribou, more recent studies (Parker, 1971) reveal fairly constant numbers.

2. Stability and Constancy of Arctic Ecosystems

Our ideas on stability and sensitivity in ecology have undergone thorough revision in recent years. The view that ecosystems with a large variety of species are particularly well buffered against environmental influences has had to be modified. Current concepts can be summed up as follows (Remmert, 1980):

1. In some ecosystems the member species fluctuate considerably about a mean value. An example of this kind of ecosystem is provided by arctic terrestrial systems. This inconstancy should not be confused with instability. Coral reefs and tropical rain-forests, on the other hand, are constant systems whose members undergo relatively small fluctuations. The large number of species in the latter is now regarded as a mechanism ensuring high turnover and a large degree of constancy of its members under continually favourable and very constant environmental conditions. Constancy of this kind should not be confused with stability.
2. Inconstant systems are for the most part resilient to ecocatastrophes and return relatively quickly to their original state.

For a tundra ecosystem it is of relatively little importance whether large-scale destruction of vascular plants is caused by a lemming explosion, by grazing reindeer or by man-driven bulldozers. The constituent species have always been subject to very large oscillations, so the system has evolved with the ability to compensate for ecocatastrophes of this kind.

Constant systems, in contrast, have not evolved the ability to stand up to such ecocatastrophes and often collapse entirely if disturbed. This is a familiar danger to the coral reefs and rain-forests of the tropics.

An additional point to be considered is that the majority of long-established systems have evolved in such a way that all utilizable matter is stored in the living substance. Nothing of any value to the organisms is left in the soil or the medium.

This is the situation in many tropical rain-forests and coral reefs. The removal of the living substance robs the system of its entire resources and consequently of the ability to regenerate. In fluctuating systems, on the other hand, only a part of the essential substances have been extracted from the medium or the soil and regeneration is therefore still possible.

Thus all arctic systems ought to be highly resilient and insensitive to human interference. To a surprisingly large degree this is so. Considering the length of time for which the Arctic has been exposed to Man and his technology it is astounding that so little evidence remains of the often drastic disturbances. About 300 years ago between 15,000 and 30,000 Dutch whale hunters invaded Spitsbergen's tundra each summer, but now almost nothing is left to remind us of their presence and the tundra has already fully reclaimed the area. The extensive settlements of the Vikings in the tundra of southern Greenland have now been completely swallowed up and no trace is detectable to the uninitiated eye although the last settlements only died out about 400 years ago. Modern mining works on Spitsbergen that were active until the 1930s are now barely detectable, the tundra having advanced again up to the very machinery. Natural ecocatastrophes such as the plagues of butterflies in Scandinavian birch forests are also rapidly made good.

The mechanism of this type of regeneration depends in part on the low degree to which members of the systems involved are bound to a particular locality. In contrast to their behaviour in temperate latitudes, species such as large owls, falcons and large mammalian predators that are usually highly sensitive to human interference show no attachment to any one locality in the Arctic: they simply depart when conditions are unfavourable and return when the situation improves. In favourable regions the rate of reproduction may be unusually high, so that losses can quickly be repaired. Furthermore, the high nutrient content of the soil remains untouched by such a catastrophe, or may even be increased (if the catastrophe involves deeper thawing of the permafrost).

Nevertheless, such reassuring remarks should not be allowed to mislead us. Northern whale populations have still not recovered from the decimations inflicted upon them by the seventeenth century whalers. Numbers sank to such low levels that regeneration will probably take a very long time. Overfishing in the region of the Barents Sea appears to have reached a point where regeneration in the near future is scarcely possible. On land, modern technical equipment represents the main source of danger to the tundra. We know that heavy machinery moving over the permafrost causes it to thaw and that this effect is long-lasting. In such places the tundra dies off, small pools form, and very probably there are additional unrecognized effects. Modern technologists have therefore aimed at causing as little disturbance to the permafrost as possible, but their efforts, in turn, have still further consequences. Roads are often constructed on supports, and pipelines laid above and not below ground, thus interrupting what were large uniform areas of tundra. The migratory routes of the animals have been disrupted and large inroads have been made on the environment of the caribou. Apparently reindeer and caribou are extremely reluctant to cross railway lines, roads or pipelines, so that the total effect of such constructions on the tundra ecosystem can be alarmingly great. This is a problem that forms the subject of large numbers of research projects at the present time.

3. The Animals in Terrestrial Ecosystems

For physiological reasons poikilothermic herbivores cannot play a large role in cold regions (p. 33). In general, therefore, we can regard their contribution to the energy cycle as minimal. This has been confirmed for Point Barrow (Alaska), Devon Island and Spitsbergen. The number of butterfly and moth species with entirely herbivorous larvae is amazingly high in the Canadian Arctic. Apparently, despite the large number of species, the numbers of individuals and their contribution to energy flow are small. The only exception is in the "warm Arctic" of northern Scandinavia where summers are in any case warm, due to the Gulf Stream, and winter temperatures never drop to very low values. In addition, there is no permafrost, and permanent ice is confined to very small areas in the mountains. Consequently, relatively high temperatures can be attained in summer, and there are vast areas of birch forest gradually giving way to pine forest towards the east. At irregular intervals, explosive increases in the numbers of butterflies (especially well known is Oporinia autumnata) cause serious damage to large expanses of the birch forests, in some cases amounting to total destruction, so that the timber line recedes (Fig. 97). In 1977 Haukioja and Niemelä reported that mechanical damage to birch leaves leads to the formation of substances poisonous to herbivorous organisms. As a result, the Oporinia populations collapse after a wave of feeding and the trees sprout normally in the following year. On the strength of such findings Haukioja and Hakala (1975) have postulated a general theory of herbivore cycles.

The second important point at which poikilothermic animals participate in the energy flow in terrestrial ecosystems is in the breakdown of litter in the soil layer. So far, little is known about the extent of their contribution. The well-known larger decomposers that play an important role in the litter of temperate regions (isopods, diplopods, earthworms and snails) are absent from the regions falling within our definition of the Arctic. The large dipteran larvae (Tipulids) are also missing. But these groups are also absent in ecosystems on acid soil in temperate latitudes (such as beech forest on granite, sand or sandstone), where their function appears to have been largely taken over by the larvae of sciarids and mycetophilids. These two groups play by far the largest role in all arctic soils (in addition to mites and Collembola). Nevertheless, even in the well-investigated regions of temperate latitudes it is still unclear whether these animals make an important contribution to energy flow or not. The significance of their mechanical action is difficult to judge: it is usually claimed that it creates favourable conditions under which bacteria and fungi can utilize the litter. There is no doubt that the two Nematocera groups play a substantial role in the energy flow in arctic soil (the only other groups of any significance are the enchytraeids), although at present it is not possible to measure their contribution. Preliminary estimates from studies made on Spitsbergen (Solem, Sendstad) suggest that the part played by piokilothermic soil animals in the breakdown of litter is a very large one, and thus comparable with the situation in more favourable climates.

Warm-blooded herbivores seem to make a relatively large contribution to the energy flow in the Arctic, similar to that seen in the steppe regions of temperate and tropical latitudes. Pöhlmann assumes that the reindeer on Spitsbergen at their

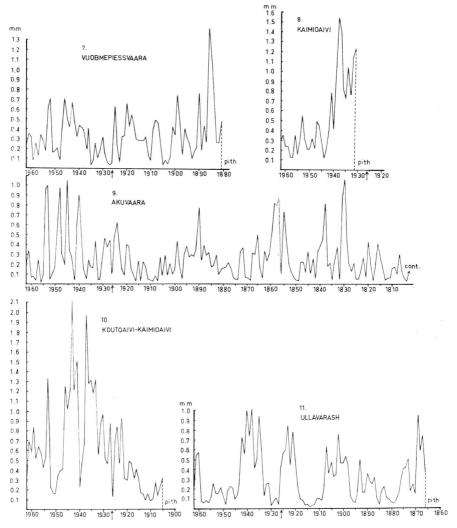

Fig. 97a. Width of annual rings in Finnish birch trees in regions afflicted by explosive increase in numbers of Oporinia autumnata. Minimum growth rates are usually connected with such outbreaks

present high ensity consume 7%–14% of the aboveground primary production. In oases in central Iceland the pink-footed goose, at a density of several pairs per ha, pecks off almost 100% of the primary production (Gardasson). But both of these examples are exceptional. In general, utilization by any one herbivorous species is less than 10%, and all herbivores taken together probably account for no more than 15%. The limitations of such an approach become very obvious when it is considered that the primary production is not equally available at all times of year, nor do the animals exploit it to the same extent throughout the year. Far more harm is caused, for example, by grazing in spring, when the plants are sprouting, than by a

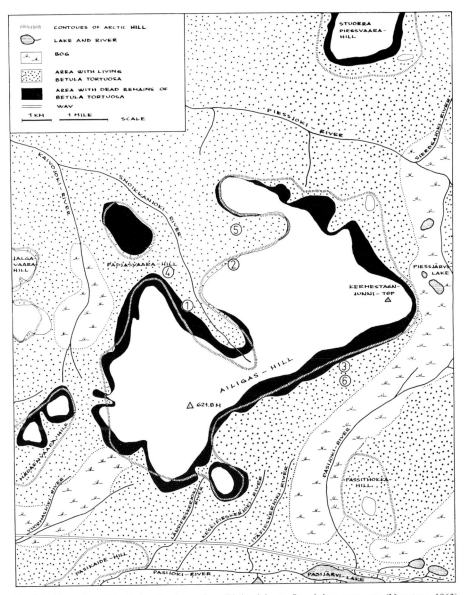

Fig. 97b. Recession of the timber line in northern Finland due to Oporinia autumnata. (Nuorteva, 1963)

similar consumption in autumn (Fig. 77). A lemming peak may result in a reduction of more than 30% in the production of the tundra, but this is due to root destruction and the quantity of plant matter actually consumed by the animals is relatively unimportant. The real effect of the selective consumption of Ericacae buds by ptarmigans (Watson and Gardasson) is difficult to estimate.

Very little is known either about the quantitative role of muskoxen and arctic hares. The former tend to prefer steeper slopes and moister areas than reindeer, and

Fig. 98. Pack formation and flight of North American arctic hares. (Burt and Grossenheider, 1964)

since they apparently never undertake migrations over greater distances it ought to be possible to obtain exact data concerning their influence. In the north of the Canadian Archipelago and in northern Greenland arctic hares tend to form packs, which have not yet been studied in any way. The fact that in the face of danger the whole pack flees, leaping like kangaroos (Fig. 98), suggests a definite social structure, but no information is as yet available concerning population densities or influence on the vegetation.

Herbivores that migrate annually and for this reason do not severely damage the sprouting vegetation in spring apparently have a higher utilization quota than stationary populations. The high degree of utilization of the vegetation by wild geese in central Iceland provides us with an example that gives cause for reflection. It may well be that goose populations of this density were much more common previously, and have been exterminated by Man.

When estimating the role of warm-blooded herbivores in energy flow it should be borne in mind that the plants are broken down much more rapidly if eaten, and their mineral components returned to the soil in readily available forms in faeces and urine. This results in an acceleration of the cycling of matter and has the same effect as manuring. The productivity of the vegetation rises and the role of the animals in the cycling process is obvious. The importance of warm-blooded animals in the cycling of matter has been pointed out by Schultz for lemmings and by Watson and Moss for Lagopus. Their data indicate that, for a large variety of reasons, the eating off of the plants leads to an increase in their mineral content. The situation is set out in detail in Figs. 99 and 66. The most important element appears to be phosphorous. Thus, herbivorous homoiothermic animal organisms play an important part in the energy flow and cycling of matter in the system as a whole, as well as exerting a considerable influence on the composition of the plant societies and overall productivity (see p. 202). The enrichment of many arctic regions by the dung of mammals and birds that seek their food in the sea has already been discussed (see p. 105).

Evidence points to the likelihood that the cycles of herbivorous birds and mammals are essential for maintaining the productivity of the tundra. This idea finds support in the studies of Bliss (1977) and his co-workers on Devon Island. The

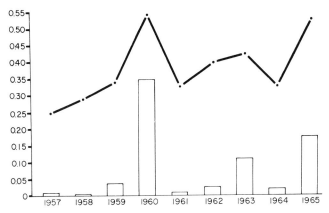

Fig. 99. Relationship between phosphorus content of plants and lemming density. (Schulz, 1972)

low numbers of lemmings and their relatively small contribution to the energy flow can be misleading: in fact their building activities are of very considerable importance to the tundra. Remineralization of dead vegetable matter is speeded up, and because the nutrients are set free more quickly plant growth is accelerated.

Little quantitative data is available concerning the influence of one animal species on the populations of another. Although it is possible to deduce information about the prey removed from a system on the basis of the food requirements of a large number of the predator species involved, this is still an unsatisfactory way of judging the influence of the predators on the populations. Birds of prey, owls, and predatory mammals, for example, exert very little influence, and probably represent no more than subsidiary factors in connection with the collapse of a population or in delaying its regeneration. On hay-fields in Iceland, Bengtson et al. (1974) recorded 238 lumbricids per m^2 on grass areas protected by nets, but only 107 on unprotected patches. The predator concerned is the golden plover, which consumed an average of 4.5 earthworms per day and m^2, or approximately 100 individuals per m^2 for the period of observation.

What the above observations have shown us is that the role of animals in the ecosystem is not confined to the steady handing on of energy and material. They have additional functions that are much more difficult to analyse, almost impossible to measure, and for the most part overlooked by modern ecosystem research. A good example is provided by pollination. About half of the vascular plants of the Arctic are flowering forms and are thus dependent upon pollination, although this does not necessarily have to take place each year: many of the plants can also reproduce by other means (parthenogenesis, self-pollination, asexual reproduction of the most varied kinds). But the formation of flowers and the large quantities of nectar produced in arctic regions are evidence that in the long run every flowering plant requires outside help with its pollination (Fig. 37). Cross-pollination has to be postulated if continual selection of genotypes is to take place to cope with changing environmental conditions. The argument that since most plants possess alternative mechanisms they are not dependent upon pollination for reproduction is unacceptable. The presence of sciarids, mycetophilids, flies (especially Syrphidae) and, in favourable regions, butterflies, moths and bumblebees as well, is thus indispensable for the existence of the arctic tundra, even if pollen transport and nectar consumption are too small to be represented quantitatively in the energy flow and cycling of matter. Worker bumblebees tend to be scarce in arctic regions (Richards, 1931) and in Bombus polaris colonies, for example, only one generation of workers is produced before sexual forms are reared. Low fecundity of the founding queen is of value in such cases as it enables a sexual brood to be produced in a short favorable season. The bumblebees of the Arctic feed on a wider variety of plants than those in temperate regions, thus ensuring that in spite of the limited season they will collect sufficient food to rear new queens. The fact that Collembola frequently visit flowers is often overlooked. It is quite usual to find a few individuals of Entomobrya comparate, for example, in the flowers of Lesquerella arctica, apparently seeking food (in the form of pollen, which they eat directly off the anthers (Kevan and Kevan, 1970)) and warmth. Similar observations have been reported for many species of Collembola, although it is much more likely that they are pollen consumers than pollinators.

The important connection between flowering plants and insects in the Arctic has been emphasized, particularly by McAlpine (1965). He demonstrated that at their imagine stage most Diptera in the region of Lake Hazen visit flowers to collect nectar (and in many cases pollen). This even holds for groups of which it has been said that they require no food as imagines. It also applies to chironomids such as the genus Smittia. McAlpine speculated that the absence of mosquitoes in some arctic regions is solely due to the lack of flowering plants (above all of Dryas integrifolia). About 80%–90% of the insects netted near Lake Hazen were carrying pollen grains, which indicates their importance in arctic ecosystems, since sooner or later pollination is undoubtedly necessary.

Similar connections, far from easy to measure, appear to exist in the case of aphids. The animals suck plant juices in many arctic regions, some from the roots in the soil and some from the aerial parts of the plants. Since these juices contain only small quantities of essential trace substances the insects have to remove far more carbohydrate-rich nutriment from the plants than they require for their metabolism. As a result they excrete a so-called "aphid sugar". Experiments have shown that the activity of nitrogen-fixing bacteria is considerably increased by the sugar-water dripping on to the ground. This is another fact to be borne in mind in any analysis of the importance of animals in the arctic tundra.

In addition to these systems and their animals there seems to exist a very different type of arctic biome, which might be called "goose-tundra" because of the vast numbers of geese of different species breeding there. This type, hitherto virtually unknown from the ecological point of view, is typical for very wet tundra reas with islands (viz. without mammalian predators) in Canada and Siberia (see p. 203 and Figs. 128, 139). Detailed studies have only been carried out in the somewhat different central icelandic colonies of pink-footed geese.

4. The Animals in Limnic Ecosystems

The dominating factors for arctic inland waters are the perpetually low temperature and the resultant permafrost. Where soil and surface freeze in winter, springs are unable to rise and streams cannot form. Those that are in existence seize up and the remaining pools freeze solid. Continuously flowing water with its source in the permafrost region is only encountered under very special conditions, namely in a river draining a larger lake. Most of these lakes receive no inflow for a large part of the year and are usually unable to sustain their outflow throughout the long winter. Nevertheless, this can happen, and an example is the Ruggels river, draining Lake Hazen on Ellesmere Island. It drops 150 m over the distance of 29 km before emptying into the sea. Despite the low winter temperatures the first 600 m always remain ice-free (Deane, 1958; Oliver and Corbet, 1966).

The amount of water carried by rivers coming from regions with no permafrost is relatively constant throughout the year. This is the case for the rivers of Iceland and northern Scandinavia, and for the great rivers of the northern Soviet Union.

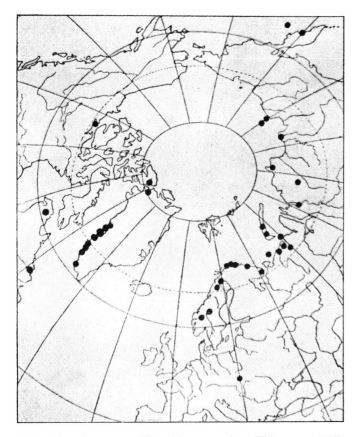

Fig. 100. The distribution of Branchinecta paludosa. (Thienemann, 1950)

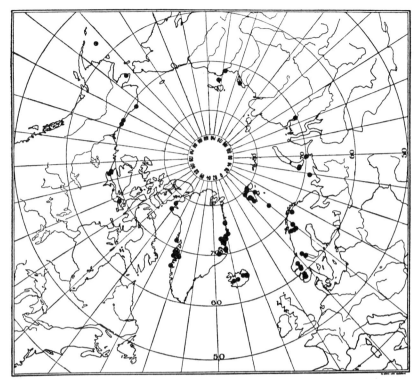

Fig. 101. The distribution of Lepidurus arcticus PALL. Its occurrence in Russia is incompletely mapped. (Thienemann, 1950)

Some of these rivers arise in relatively warm regions and receive water with spring-like temperatures at a time when the winter cold on the coasts of the Arctic Ocean is at its most severe. Long after the thaw has set in in the upper reaches of such rivers the tundra is still frozen and the river bed is still completely encased by ice, so that it cannot expand to accommodate the much larger quantities of water. As the water velocity increases the ice may burst with explosive force. (No up-to-date studies have been carried out on the fauna of such rivers. The effect of the extreme pressure changes in spring, especially on fish with a swim-bladder, ought to be investigated.)

In addition to dictating the behaviour of flowing water, permafrost also has an influence on stagnant water in the Arctic. When the ice and snow on the permafrost melt, innumerable small pools form. This was also described for regions without permafrost by Thienemann in 1938 (Figs. 100, 101). Thus ideal conditions are provided for insects whose larvae live in water, which explains the enormous numbers of mosquitoes in northern forests and tundra. The smallest of the pools dry out relatively soon, whilst water drains from the larger ones via outflows. What remains is a myriad of small to large, shallow ponds, whose water level sinks considerably over the course of the summer. At the same time the pH of the water is displaced from an acid to a more basic range. Oliver and Corbet (1966) assume two reasons for this: firstly, that photosynthesis deprives the water of CO_2, which leads to a rise in pH. A sudden drop in the pH at the end of the growth period presumably marks the cessation of algal photosynthesis. Concurrently with the rise in pH in summer the conductance of the pond water also increases. Like the change in pH, this is particularly noticeable in regions with no precipitation and in ponds with no appreciable outflow. Oliver and Corbet consider the rise in conductance to be due to the drop in water level over the summer, and to the thawing of the permafrost. The latter releases more and more ions, which remain in the water. Such phenomena are less obvious in ponds with considerable inflow and outflow and in those in regions with precipitation, as well as in ponds on very poor geological formations, where very little dissolved material can enter (Fig. 102).

The temperature of small bodies of water of this kind usually reflects the air temperature, although there are certain exceptions. In very exposed regions with low air temperatures the water temperature may be relatively high (e.g., on Spitsbergen). Rakusa reported such a situation on the Hornsund in southern Spitsbergen, where the mean temperature of small bodies of water in August and September was on an average $2°–4°$ C higher than the mean air temperature. The water temperature in ponds in wind-sheltered localities (common in the Canadian Arctic) is usually only slightly below that of the air. The daily temperature fluctuations of ponds in the High Arctic are often much more pronounced than on dry land, which may account for the relatively rapid growth of their inhabitants.

Larger arctic lakes are found mainly in northern Scandinavia, Finland, immediately south of the Arctic Circle, in Iceland and in the Canadian Arctic. The Fennoscandian and Icelandic lakes are situated in comparatively warm regions outside the permafrost zone, whereas those in the Canadian Arctic are within the typically cold climatic region with permafrost.

These lakes represent a very special problem. They can only have become filled with water when the ice receded following the glacial epoch, but the region had permafrost even before the lakes existed. That such lakes can exist at all, and are not simply solid blocks of ice, is not at first explicable. It seems impossible, or at least very unlikely, that water could remain unfrozen in winter in arctic permafrost regions, with the exception of inflow from warmer regions. But such lakes do exist, and although they are ice-covered for 10–12 months of the year the ice is relatively thin (up to 4 m). Even relatively shallow lakes on Spitsbergen can provide the population with drinking water all the year round, although quantities are sometimes limited. Receding glaciers in the Antarctic revealed a lake which nowadays harbours blue algae and bdelloid rotifers. It is perpetually covered by several metres of ice, and at no place on the entire lake does the ice ever melt. It seems that lakes of this kind act as solar energy traps: the sunlight is refracted on the rough surface of the ice into the lake, and penetrates to fair depths which warm up more than the cold, wind-exposed surface. The process of warming-up initiated beneath the surface ice eventually causes the lower ice layers to melt. Therefore the existence of ice-covered arctic lakes containing unfrozen water is theoretically possible.

So far, very little information is available concerning the animal life in arctic inland waters. In particular, we know relatively little about specific mechanisms of adaptation, or about the populations inhabiting different types of running water. Little can be said as to which kind of smaller body of water freezes solid in winter

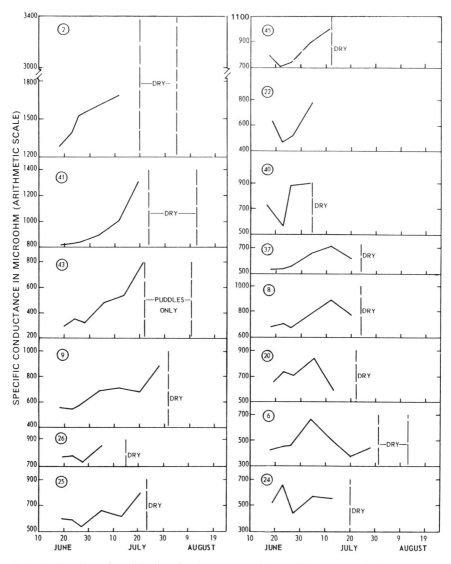

Fig. 102. Chemistry of small bodies of water near Lake Hazen, Ellesmere Island, during one growth period. (After Oliver and Corbet, 1966)

and which kind still contains water, whilst the chemistry of these persistent bodies of water, particularly their oxygen content, is an almost untouched subject.

It is surprising that the animals of the mountain streams of temperate latitudes are rarely encountered in what we define as the Arctic. Even relatively large arctic rivers, if they are non-existent in winter, apparently harbour only chironomids of various genera. Characteristic species of temperate mountain streams, such as Plecoptera, Ephemeroptera, Amphipoda and snails, for which the arctic streams would seem to be predestined on account of their temperature, are usually as rare as typical mountain-stream fish like Salmo, Lota and Cottus, or even missing. All of these forms seem to require constantly flowing water, and are thus not found in streams on genuine permafrost. They only penetrate far north of the Arctic Circle on the Scandinavian peninsula and in Finland (together with some Trichoptera) where practically all waters flow throughout the year. The only fish that appears to be of any importance in the Canadian Arctic is Salvelinus, although it is in fact a species of the large lakes and only enters streams for short periods of time to spawn (and not every population does so). Two other fish species found here are the three-spined stickleback (Gasterosteus aculeatus) and the nine-spined form (Pungitius pungitius).

Very few studies have been made on the lamellibranchs of arctic lakes. Only representatives of the Sphaeriidae family occur, but they penetrate even to High-Arctic (and High-Alpine) lakes. The filtration values of Pisidium vary between 3 and 7 ml/g live weight (inclusive of shell and mantle cavity water) an hour. Assuming a mean density of 500 animals/m² in two summer months, with a probable water temperature of 11°–13° C, a water layer of the order of ½–3 cm thickness can be filtered. At first sight this does not seem to be very much, until we reflect that filtration is accompanied by crawling and digging activities on the interface between water body and substrate (Hinz, 1976). Further studies are urgently required: there is no valid reason for ignoring this animal group in future analyses of limnic systems. There is also an urgent need for a comparative analysis of the individual biological components of a lake containing relic marine forms and one without. The interspecific relationship of the fauna of arctic freshwaters are so far virtually uninvestigated. High-Arctic Daphnia, for example, exhibit no diurnal vertical migrations and are almost unpigmented and transparent, whereas the more southern forms, which do migrate vertically, are strongly pigmented. Is this connected with the danger of predatory fish, or are other factors responsible? This state of ignorance is reflected in the lack of clear-cut qualitative food analyses for High-Arctic fish, and the fact that it has so far been impossible to construct a realistic energy-flow diagram.

Eutrophication represents a special problem in High-Arctic lakes. The long period for which the lakes are cut off from the air by ice renders eutrophication a greater danger than in lakes of warmer regions. there are long periods of time during which oxygen is merely consumed and productivity is extraordinarily low, so that the lake has a negative oxygen balance.

5. The Animals in Marine Ecosystems

Conditions in the oceans are dictated by the currents. The relatively warm water along the west coast of Scandinavia as far north as Spitsbergen (and on the west coasts of Canada and Alaska) is brought by currents flowing from the south, whilst cold currents flowing from the north along the east coast of Greenland are responsible for the considerable southerly displacement of the edge of the ice (Figs. 103, 104, 144). This combination of oceanic currents results in nutrient-rich regions

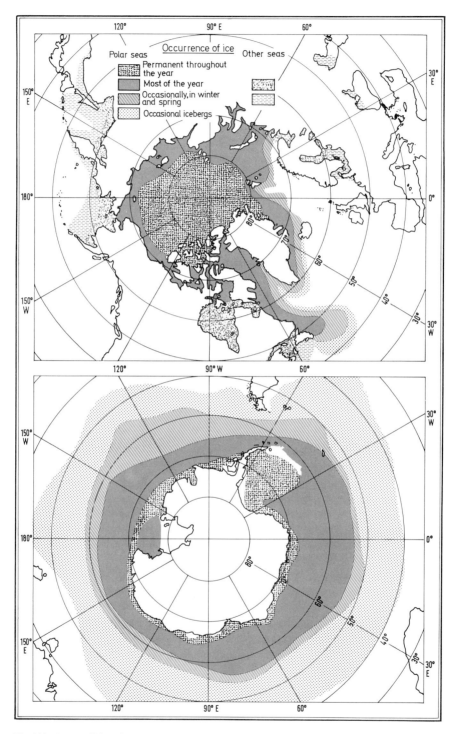

Fig. 103. Ice conditions in the Arctic (*above*) and Antarctic (*below*). (After Dietrich et al., 1975)

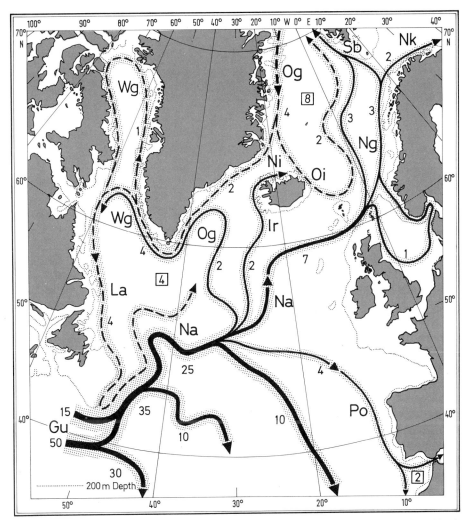

Fig. 104. Scheme of the water transport (in 10^6 m^3 s^{-1}) in the 0–1000 m layer in the northern North Atlantic Ocean (adapted from Wegner, 1972). *Gu* Gulf Stream; *Ir* Irminger current; *La* Labrador current; *Na* North Atlantic current; *Ng* Norwegian current; *Nk* North Cape current; *Ni* North Ireland current; *Og* East Greenland current; *Oi* East Ireland current; *Po* Portugal current; *Sb* Spitsbergen current; *Wg* West Greenland current, *Squares* water sinking in 10^6 m^3 s^{-1}. *Solid lines* relatively warm currents. *Interrupted lines* relatively cold currents. (After Dietrich et al., 1975)

which are still further enriched by inflowing Siberian waters. The result is a high primary production in many places and very large numbers of animal consumers (see also p. 214f.).

In most of the Arctic, however, this situation is not immediately obvious. The marine littoral seems to be almost devoid of animal and plant life, in contrast to the characteristic richness of the corresponding zone in temperate and tropical latitudes. The poverty of the arctic littoral is a consequence of drifting ice. Almost all coastal regions in the Arctic are exposed to the influence of drift ice and are blocked in winter by pack ice, which also has a highly erosive effect in connection with tidal movement. Living organisms thus have little chance of developing, although the fact that sea ice is as a rule only about 80 cm thick means that light can pass through and photosynthesis can take place in the underlying water. This explains the presence, even in the High Arctic, of a rich marine flora of large seaweeds (Laminaria) with a corresponding fauna, immediately below the tidal region.

Drift ice and frost are also important factors for the littoral fauna on the Baltic sea coasts, the North Sea coast of Germany, and on the eastern coasts of North America. A striking example of the way in which an animal of the tidal zone (Littorina saxatilis) can adapt to very cold conditions when exposed at low tide was reported by Kanwisher (1955, 1959). At $-15\,°C$ more than 65% of the snail's body water is frozen, but the ice is exclusively extracellular. The body cells are dehydrated and shrink, and drastic alterations in the cellular membrane system take place. The cells themselves, however, never freeze at this temperature. No similar studies have been made in genuine arctic regions, although the situation is probably very similar in the few animals of the tidal regions.

Northern Scandinavia as a constantly ice-free region does not fit into this general picture at all. The warming influence of the Gulf Stream makes itself particularly noticeable here (even as far as Spitsbergen, Figs. 105, 106, 144), with the result that the littoral fauna and flora are in every way comparable with those around the British Isles. In sheltered localities the rocks are lavishly covered with Fucus and the fronds of Ascophyllum, and in the deeper water vast areas of Laminaria harbour a rich marine littoral fauna. Outposts of this rich littoral flora and fauna can be found as far north as Spitsbergen, whose west coasts remain free of ice in some winters: dense patches of Fucus can even be found in places (e.g., Magdalena Bay) and a small amount of Balanus and Littorina saxatilis is regularly present in the tidal zone.

This is perhaps an appropriate point at which to put the question as to the significance of the Arctic Circle in the polar seas. Neumann found that moonlight, the zeitgeber for the synchronization of reproduction in marine animals, becomes less and less effective towards the north. The limit of its influence is at a latitude of approximately 60° N, above which the summer nights are too short and too light. Although animals still exhibit a lunar rhythmicity on Helgoland it is no longer governed mainly by the moonlight (at least in Clunio) but by the turbulence connected with the tidal hub. We can therefore assume that a physiological limit is attained somewhere between 55° and 65°N. Whether all organisms switch over, like Clunio, to other zeitgebers, and whether the synchronization of reproduction in northerly latitudes is achieved by other means than in more southerly regions is so far unknown. In any case, to the north of this line moonlight itself cannot act as a zeitgeber. Light is obviously an important factor for the marine fauna, too, and consequently the Arctic Circle is a valuable means of indicating the threshold to the Arctic, although much remains to be done in this direction.

According to Thorson (in Hedgepeth, 1957), all marine benthic communities belong to the complex of Macoma communities. The principal arctic variant is formed by Macoma calcarea, Cardium ciliatum and Mya truncata (Fig. 107) and is thus very similar to other Macoma communities frequently encountered on the west

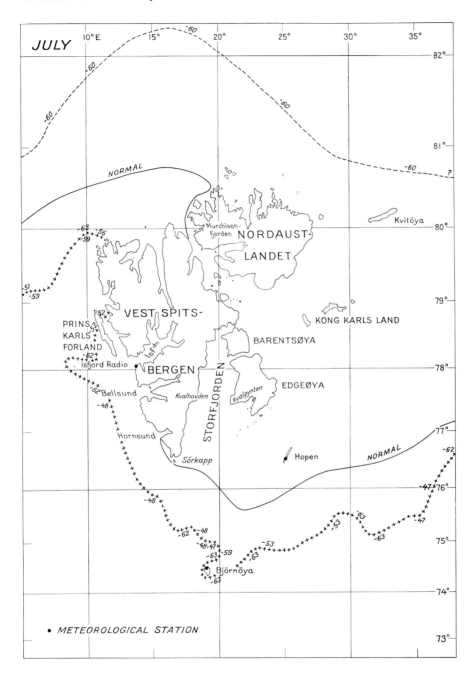

Fig. 105. Mean July ice limit near Spitsbergen for July, 1945–1963. Extreme positions of the ice edge in the years indicated are also given. (Lunde, 1963)

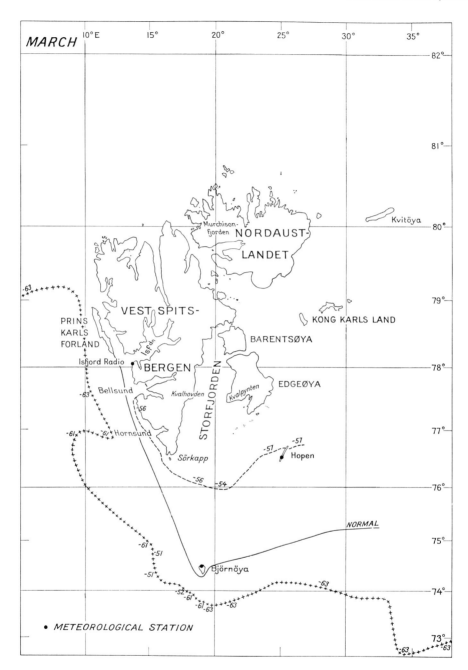

Fig. 106. Mean ice limit near Spitsbergen for March, 1946–1963. Extreme positions of the ice edge are indicated and the year of occurrence. (Lunde, 1963)

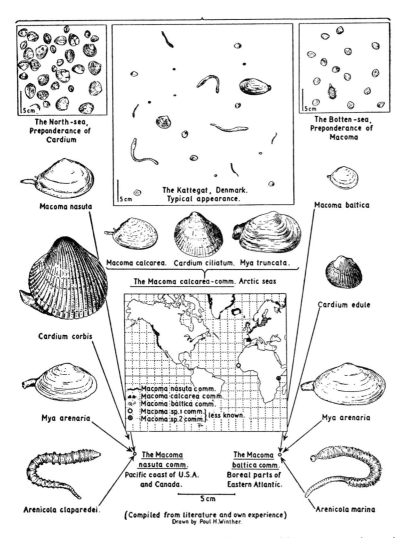

Fig. 107. Oceanic regions in the Arctic mainly belong to a Macoma community, such as is typical of boreal shallow seas. (Thorson in Hedgepeth, 1957)

coast of the United States or in the boreal parts of the eastern Atlantic. The lamellibranches (species by no means confined to the Arctic) provide excellent food for fish (cod and halibut) and seals, and in fact constitute their basic diet. It can be seen from Fig. 108a and b that considerable deviations from this scheme are possible. Brittle stars and Crustaceae may be dominant in some places, or the lamellibranch Astarte borealis and polychaetes in others. The Gulf of Bothnia, which in some places exhibits almost genuine arctic conditions, can be regarded as still belonging to this complex. It is in fact a brackish arctic lake with a number of relic glacial forms such as Macoma, Mesidothea, Pontoporeia and, in places, Astarte (Fig. 109).

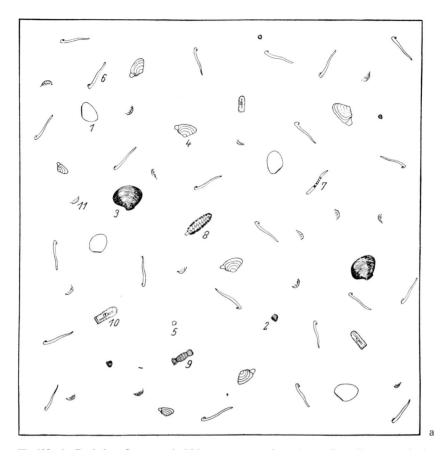

Fig. 108a, b. Deviations from a typical Macoma community. **a** Astarte borealis community in north-west Greenland (at a depth of 14 m). *1* Macoma calcarea; *2* Astarte (Nicania) banksi; *3* Astarte borealis; *4* Mya truncata; *5* Thyasira flexuosa; *6* Terebellides stroemi; *7* Maldanidae; *8* Harmothoë imbricata; *9* Priapulus caudatus; *10* Myriotrochus rinki; *11* Amphipoda **b** Chiridothea community in north-west Greenland (at a depth of 23 m). *1* Mesidotea (Chimdothea) sabim; *2* Diastylis rathkei; *3* Amphipodae; *4* Ophiura sarsi; *5* Pectinaria granulata; *6* Myriotrochus rinki; *7* Thyasira flexuosa; *8* Astarte (Nicania) banksi; *9* Mya truncata; *10* Cardium groenlandicum. (Vibe, 1939 from Remane, 1940)

The marine benthic communities of the Arctic thus fit well into the general picture, with little to suggest a sharp distinction from marine benthic communities of more southerly regions, although some typically arctic species do occur. In addition, there are certain striking differences between warmer and cooler seas with respect to the relative abundance of the various groups (Figs. 110–115). Attention was drawn to this point, chiefly by Thorson (in Hedgepeth, 1957). He showed, for example, that there is a striking decrease in opistobranchs from tropical to arctic regions, whereas the numbers of Cumaceae and Holothuria remain almost constant. He also pointed out that it is particularly the benthic forms (epifauna) that decrease in arctic regions, whilst forms living in the bottom itself (in-fauna) are involved to a lesser extent. At present we can only speculate as to the reasons for this. Again, with

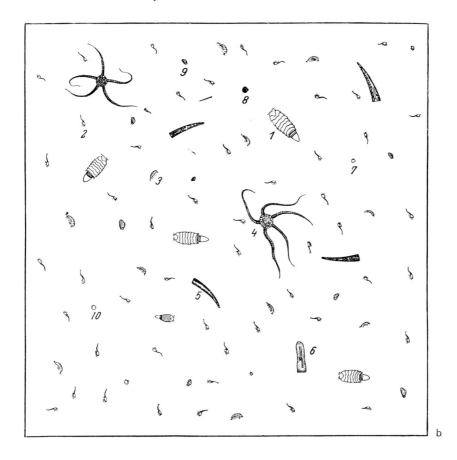

b

increasing latitude, more and more benthic forms with planktonic larvae are missing. They are replaced by increasing numbers of forms with direct development or greatly reduced larval periods, and brood care is also more common. So far it is impossible to say which of these forms of reproduction costs the most energy. It is possible that the reduction in the enormous number of planktonic larvae results in a saving of energy for the adult animals and that the available energy is invested in growth. Whilst a number of animal groups develop giant forms in arctic and antarctic regions, others are dwarfed. Arnaud (in Llano, 1977) arrives at the conclusion that all forms requiring larger quantities of calcium for growth are dwarfed because precipitation of calcium from the water at very low temperatures is extremely unlikely, in contrast to the situation in tropical waters. This category includes lamellibranchs, snails, echinoderms and many Bryozoa. With increasing cold the shells of such animals gradually become more fragile and lighter. On the other hand, forms that tend to gigantism are those whose growth is unhindered by lack of calcium (crustaceans, polychaetes, opisthobranchs, Ascidia and hydroids).

Data concerning the corresponding microfauna are not available. It must be of significance for land and freshwater regions that the tidal regions, except along rocky coasts, are inhabited by a very rich microfauna, even as far north as the High

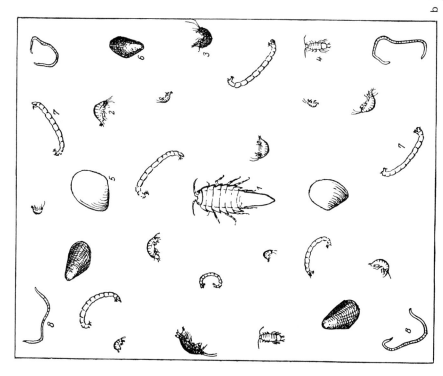
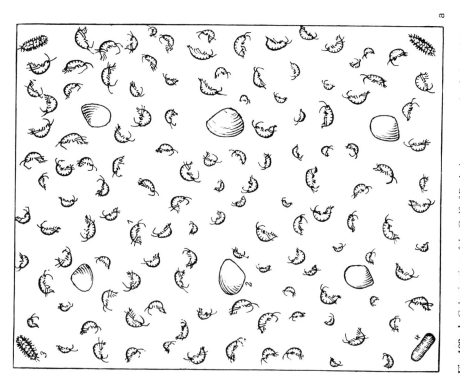

Fig. 109a, b. Colonization of the Gulf of Bothnia as an example of a brackish region of the subarctic waters. **a** Pontoporeia colony. *1* Pontoporeia affinis; *2* Macoma baltica; *3* Harmothoë sarsi; *4* Halicryptus spinulosus. **b** Mesidothea colony. *1* Mesidotea entomon; *2* Pontoporeia affinis; *3* Gammarus locusta; *4* Asellus aquaticus; *5* Macoma baltica; *6* Mytilus edulis; *7* Chironomid larvae; *8* Tubifex tubifex. (Segerstrale, 1973 from Remane, 1940)

The Animals in Marine Ecosystems

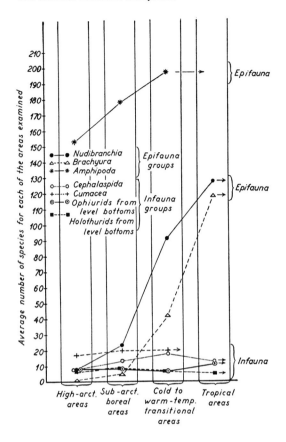

Fig. 110. Changing composition of the benthic fauna from the tropics to arctic regions. (Thorson, 1952 in Hedgepeth, 1957)

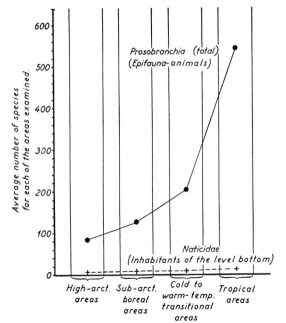

Fig. 111. Average number of species of marine prosobranchs from the tropics to the Arctic. (Thorson in Hedgepeth, 1957)

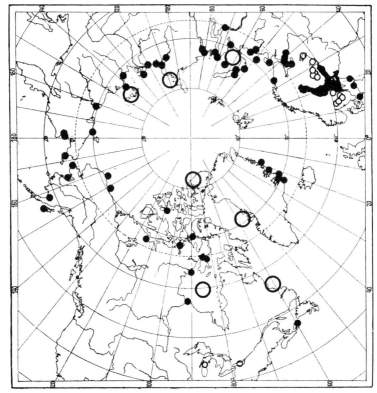

Fig. 113. Distribution of Cottus quadricornis in the Arctic. (Ekmann, 1953)

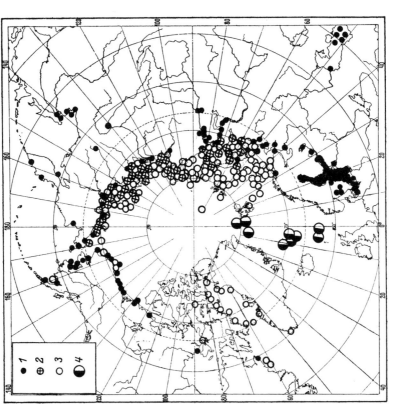

Fig. 112. Distribution of the genus Mesidothea. (Ekmann, 1953)

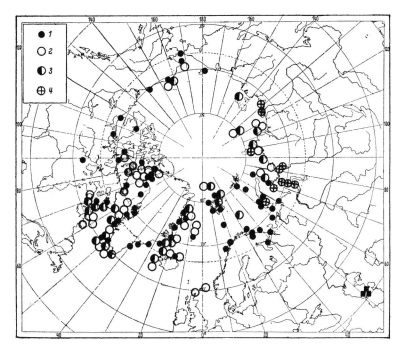

Fig. 114. Distribution of the amphipod genus Pseudalibrotus. (Ekmann, 1953)

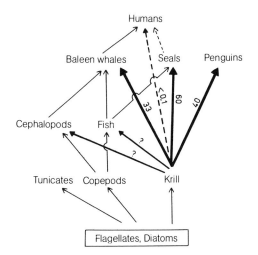

Fig. 115. Position of krill and homeotherms in the food chain in Antarctic waters. Homeotherms are the main consumers of the production. The same probably applies in parts of the Arctic. (After Hempel, 1977)

Arctic, i.e., in regions where drift ice plays an important role. This microfauna does not differ basically from that of warmer regions, although copepods and turbellarians are rarer than in southern latitudes, so that the nematodes play an even greater role (Gerlach, 1965).

Few attempts have been made to estimate the energy flow in arctic seas. In all probability, as in the Antarctic, warm-blooded animals play the predominant role.

Fig. 116. Arctic distribution of the bearded seal, a form that stays near land, and the pelagic hooded seal. *Black areas*: where cubs are born; *dotted line:* drift ice limit at the time when the cubs are born. (After Van der Brink, 1957)

Seals and penguins appear to have increased in numbers following the decimation of Baleen whales in the Antarctic, and even attain sexual maturity at an earlier age nowadays (Hempel, 1977). Although both whale and seal populations have been reduced to a far greater extent in the Arctic than in the Antarctic, it can be assumed that the role played by these mammals and the birds is (at least potentially) dominant in the Arctic just as in the Antarctic. The importance of seals in the oceans can be judged from the fact that many of them never come near land (hooded seal and harp seal). They mate and breed on large ice floes, the mating season following immediately on the birth of the young (Figs. 76, 116; Table 4).

V. Types of Arctic Climates

As we move southwards through the Arctic the daily temperature fluctuations increase in amplitude (Figs. 117, 118), a fact that is of considerable biological importance, especially with respect to decomposition. In addition, just as in tropical and temperate climates, we can distinguish between two continuously connected extremes of climate: the dry arctic climate (continental arctic climate) on the one hand, and the wet arctic (oceanic arctic) on the other. What does this mean?

If large quantities of precipitation fall during the long arctic winter the snow cannot melt in the short cool arctic summer and a continual blanket of snow and ice results. Further, solar radiation is reflected instead of being absorbed by the soil, and thus a regulatory circuit with positive feedback is established, the mean temperature drops and a situation results in which active life is impossible. At the other extreme, where precipitation is entirely lacking, the biological end result is the same. In the absence of both snow and ice the dark ground can absorb heat throughout the summer and can become quite warm. Despite this, life in this theoretical type of climate is also impossible because there is no water. However, only the former of these two extremes has developed, and the latter exists only in a

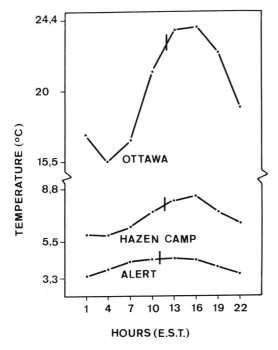

Fig. 117. Daily temperature pattern at the end of June in Canada. The diurnal fluctuations become much larger towards the south

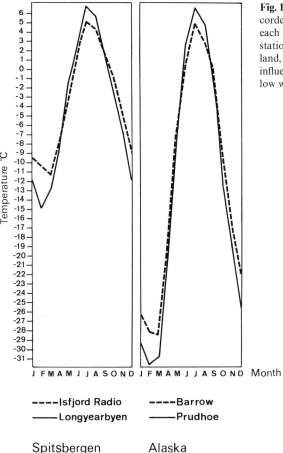

Fig. 118. Annual temperature pattern recorded on Spitsbergen and in Alaska. In each case recordings were made at an arctic station near to and strongly influenced by land, and at another under strong oceanic influence. The marine station clearly has fewer low winter and fewer high summer values

mitigated form. Apparently the presence of a perpetual ice covering is often not so much a question of prevailing temperature as one of quantity of precipitation. If we consider a section passing through Norway, far south of the Arctic at approximately the latitude of Bergen (60°N), where the coastal regions receive large amounts of precipitation, we find very large glaciers at low altitudes, as well as the vast arctic-like region known as Hardangervidda. Further east there is less precipitation, snowline and treeline are much higher and the Dovre-Fjell gives a less arctic impression than the Hardangervidda. Climatologists are agreed that larger amounts of precipitation on Spitsbergen would result in complete glaciation. This would apply equally to the islands of the Canadian Archipelago and to Peary Land in the north of Greenland. Such dry regions are of course supplied with water from surrounding areas during the summer, and are never completely without water. Life can develop because they are relatively warm in summer. Similar regions form oases in the inland ice in the Antarctic as well.

The situation is represented schematically in Fig. 119. Dry arctic climates are characterized by severe winters and warm summers, whereas the wet arctic type of

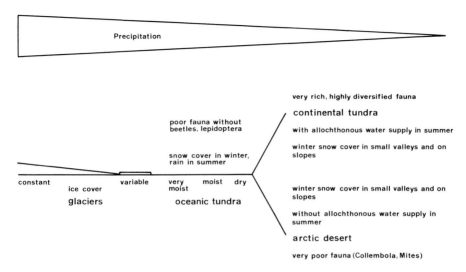

Fig. 119. Schematic representation of the dependence of tundra forms on the quantity of precipitation. At very high levels of precipitation the region is covered by permanent ice; where precipitation is less, there is first wet, then dry tundra and finally arctic desert. Where the latter receives allochthonous water during the summer months it is replaced by a rich vegetation and fauna. The highest primary production is found in the oceanic tundra, where water is not a minimum factor. Primary production in the continental tundra is almost entirely confined to the immediate vicinity of water. With the same mean annual temperature there many even grow taiga. (cf. Lieth and Whittaker p. 239)

climate has mild winters and relatively cool summers. Fog and rain are fairly frequent in the wet arctic climate; extreme examples are provided by Bear Island and Jan Mayen Island. The advantage of a wet arctic climate is that water is never a limiting minimum factor, whereas the dry type of climate has the advantage of higher summer temperatures. The disadvantage of the former is that the relatively long periods of fog and rain mean less warmth for the warm-blooded animals, whilst in the dry type of climate the meagre snow cover in winter is a disadvantage for the vegetation, which is afforded little protection from the very low temperatures. Bearing these theoretical remarks in mind we can now take another look at Fig. 119. Hardangervidda has an annual precipitation of about 1000 mm, the west coast of northern Norway receives about 500 mm, the greater part of the Canadian Arctic about 250 mm and the northern part of the Canadian Archipelago and northern Greenland 100 mm at the most. If these figures are compared with our predictions about climate we would expect to find that northern Greenland and the northernmost regions of the Canadian Archipelago still have a relatively rich flora and fauna. A glance at the distribution of arctic mammals shows clearly that this is so. Reindeer, muskoxen, lemmings, weasels and wolves are found barely 800 km from the pole at the northern tip of Greenland. The topography of the region is also partially responsible for the rich flora, in that it is hilly to mountainous, and what little snow there is can collect in the valleys and hollows and afford protection to the plants. The greater part of the land remains free of snow and warms up to such an extent in summer that the snow even melts in the hollows with their rich flora and fauna. Adequate supplies of water are also available in summer from the wetter regions farther south, and at least along streams and rivers a rich flora can develop.

Although the poikilothermic and homeothermic fauna of the continental tundra of the northern Canadian Archipelago and Greenland is of great richness and diversity, its primary production is very small in comparison. The limiting factor for the plants is water, and only in its vicinity does a closed covering of vegetation develop (the normal situation in the oceanic tundra). Thus, although the flora of the continental tundra is just as diverse as that of the oceanic tundra, its productivity is very much lower.

It appears that dry arctic regions that are supplied with water from wet arctic regions, and are topographically suited to supporting a flora, offer better conditions for living organisms than the wet regions. This, too, can be explained theoretically. Dry arctic regions with somewhat higher temperatures provide better living conditions for poikilothermic animals than the wet regions, and the mean size of these poikilotherms is greater than in the latter. Therefore in the dry Arctic butterflies, moths, and bumble-bees are more likely to be encountered and the flowers more likely to produce nectar than in the wet Arctic (see Fig. 37), which implies a greater diversity in the flora. A corresponding situation is seen with the fauna. Muskoxen seem to be extraordinarily sensitive to the wet arctic climate, and all attempts to introduce them into a wet oceanic climate have been a failure (northern Norway, Iceland). Although they thrived well at first on Spitsbergen, and even multiplied, the population has now collapsed, probably as a result of a succession of severely oceanic years with rain throughout the winter. The very dry Dovre-Fjell in Norway appears to suit them admirably. Reindeer are apparently less sensitive but they, too, appear to have died out in the very wet parts of Iceland into which they had been introduced, and have only survived in the relatively continental central regions. Thus, as compared with wet arctic regions, the warm-blooded fauna in the dry parts of the Arctic shows greater diversity, although the smaller contribution of marine birds and mammals to the fauna and the consequent reduction in amount of dung results in a certain degree of impoverishment.

What has so far been said provides a very rough survey of the situation, which will now have to be elaborated upon. To begin with, it can be said that in general the climate is wetter in the vicinity of open water than farther inland. The expression "open water" is important here because the ocean surrounding the northern parts of the Canadian Archipelago and around northern Greenland is perpetually covered with pack ice, with the result that the climate is as dry and continental as farther inland. Particularly wet arctic regions are therefore confined to a narrow strip along the coasts of mainland and islands, whereas inland and in the central parts of the islands the climate is more continental. A classical example of such conditions is seen on Spitsbergen (Svalbard), as was described by Sumerhayes and Elton (in Remmert, 1966) and confirmed in recent studies (Fig. 129). The central part of Spitsbergen has much higher summer temperatures (and much lower winter temperatures) than the western coastal regions and receives much less precipitation, with the result that it has a particularly rich fauna and flora (Fig. 119). Betula nana and a number of other warmth-requiring plants grow only in the central regions (see Rønning, 1964). Only here are mosquitoes (Aedes), herbivorous sawflies, beetles

Table 9. Comparison between species numbers of invertebrates in various arctic terrestrial localities

	Moor House England 54°65' N	Hardangervidda Norway 60°36' N	Messaure Sweden 67° N	Iceland 65° N	Barrow Alaska 71°21' N	Lake Hazen Canada 81°49' N	Spitsbergen 78° N
Collembola	56	33	?	58	32	14	40
Ephemeroptera	12	3	22	1	0	0	—
Odonata	2	0	11	—	0	0	—
Orthoptera	1	1	3	—	0	0	—
Hemiptera	71	12	?	13+	2	4	4
Neuroptera	3	1	16	1	0	0	—
Coleoptera	115	74+	900+	170	12	4	3
Trichoptera	31	3	147	10	5	1	—
Lepidoptera	70	48	170+ [a]	50	8	19	1 (?)
Diptera	453	84	?	300+	120	142	100+
Hymenoptera	85	35+	?	200+	30	57	19+

[a] +Microlepidoptera to be added

and the occasional moth (Plutella maculipennis) to be encountered, and only here have muskoxen been able to survive. The same situation is seen on Iceland and Greenland, where the terrestrial fauna and flora become increasingly rich from the coast towards the central regions. (Differences of this nature are modulated by the soil conditions, which should be kept in mind.)

We are now in a position to construct a simplified scheme of the animal life of the arctic tundra (Table 9). The number of animal species and groups is low, but these few form the basis on which large and very complex tundra ecosystems can be maintained. Breakdown of litter is carried out entirely by Diptera larvae, Collembola, mites and enchytraeids. Predatory spiders (linyphiids and micryphantids) and parasitic Hymenoptera, chiefly Ichneumonidae, exert a regulatory influence on the populations. Of the herbivores, only the homeotherms are of any importance (reindeer, muskoxen, arctic hares, voles, geese and Lagopus), of which some reindeer populations (Spitsbergen, Peary Land), muskoxen and arctic hares are relatively stationary, whilst the rest migrate over large distances. Predators (snowy owl, skua, weasel, wolverine, wolf and falcon) put in erratic appearances depending upon the occurrence of their prey, and are rarely bound to one particular locality. Diptera of a wide variety of groups, possibly also Collembola, are the most likely pollinators, with butterflies, moths, hover-flies and bumble-bees in continental (dry) regions.

The combination of the short growth period, the undoubtedly high losses suffered during the winter and the unreliability of the time at which the thaw sets in, favour a typical r-selection (see Remmert, 1978), since the animals have to be able to reproduce very quickly. This explains why the number of parthenogenetic species in many groups, especially the Diptera, increases towards the north. Since the early investigations that revealed a similar state of affairs in isopods (although far south of the Arctic, between the temperate and boreal zones), no further investigations have been carried out. The number of micropteric and even apteric insects also shows an increase as compared with boreal and temperate latitudes. This is very marked in certain Nematocera and ichneumonids, although no figures are available. And finally, the mean size of spiders and winged insects drops considerably (Fig. 29). Since the growth period lasts only one month, and at the same time is very cold, large insects would require many years to complete their development, so that small insects are quite obviously at a selective advantage.

Wet arctic regions support deciduous forests (chiefly birch) and their tundra is composed of grass and moss, whilst the dry arctic regions may support a taiga forest and a tundra vegetation consisting mainly of a variety of Ericaceae and Dryas. Lichens are found everywhere in the forests and tundra of the dry arctic zone. Pine (Pinus) and Ericaceae are almost universally considered to be indicators of fire-prone environments, and this is the case in the dry taiga and tundra. In the north of Sweden and Finland the taiga is largely devoid of spruce because it is more sensitive to fire than the pine. Only after appropriate measures were taken did the numbers of spruce (Picea) begin to increase (Zackrisson, 1977). The dry tundra regions covered by Ericaceae are also regularly exposed to natural fires.

Arctic climates are characterized by their unreliability. This is expressed in the relatively large fluctuations in mean annual temperatures as compared with temperate and tropical latitudes. Even greater differences are revealed if the mean temperatures for the month of July, which is the month mainly responsible for growth of vegetation, are compared for all three latitudes . In view of the very large fluctuations it can be concluded that the time required for the development of poikilothermic animals in the Arctic cannot be predicted with the same degree of certainty as in more favourable regions. Some species probably delay their hatching from a cold summer to the following year. Little is known about this, but since the pupae of certain butterflies and moths in temperate latitudes can remain in the same state for several years it seems safe to assume that the same phenomenon may be developed to a still larger extent in the Arctic.

VI. Case Studies

Investigations on the ecology of arctic animals have by no means covered the entire geopraphical area. In fact, the literature on the Arctic repeatedly refers to the same small number of place names. All of the data presented here originate from these regions, i.e., Spitsbergen (Svalbard), northern Scandinavia (Abisko, Kevo, Messaure), Point Barrow and Prudhoe in Alaska, Ellesmere Island and Devon Island in the Canadian Archipelago. Additional investigations have been carried out in Scoresby Sound in eastern Greenland, and a number of studies originate from regions south of the Arctic Circle, such as Hudson Bay, south and west Greenland, Iceland (Myvatn, Thjorsaver) as well as from Hardangervidda and Dovre-Fjell in southern Norway. Other regions are poorly represented: examples are provided by the Sarek region of Sweden, where faunistic studies got off to such a good start 50 years ago, and the important region of northern Greenland. The significance of the better investigated of these areas can best be illustrated in the form of a series of case studies

1. "Warm" Arctic: A Section Through Northern Scandinavia from Tromsø (Norway) to Kevo (Finland). (Fig. 120)

Scandinavia is the only arctic region without permafrost. Its coast is ice-free throughout the entire year, the growth period is relatively long and summer temperatures are high. On the Norwegian coast the climate is maritime, which in combination with the strong coastal winds is responsible for the absence of trees right down to sea level. At the inland ends of the long fjords the timberline ascends to 300 m above sea level and the forest consists chiefly of birch with a slight scattering of pine. There are still orchards and vegetable gardens (currants and rhubarb) as well as cattle farming with hay harvesting. Compared with the coast of southern Norway rainfall is low, with barely 500 m annually. Consequently there is considerable incoming radiation even at 300 km north of the Arctic Circle. The proximity of the sea ensures a long period of growth.

To the east of the steep mountains which provide shelter from wind and weather the climate is more continental, the growth period shorter, precipitation even less, as a consequence of which there is stronger sunshine. Due to the altitude and the short summer season neither cattle farming nor horticulture are possible. The valleys are covered with birch forest, the proportion of pine increasing towards the east until in Kevo it is the predominant species. We have now reached the very continental region with its brief growth period, where temperatures are high in

Fig. 120a-d. Types of landscape in northern Scandinavia. **a** Near Tromsö (Norway) **b** Abisko (Sweden) **c** Finland **d** for comparison: East Greenland tundra near the polar circle

summer and low in winter. Only a few metres in altitude separate the pine forests from the treeless Fjell plateau.

Within a relatively short distance we have thus left the temperate and fairly warm regions with their long growing season, and have entered the dry continental regions where growth is confined to a very short period, but with very high summer temperatures.

This is a situation that is thoroughly atypical for the Arctic and, not surprisingly, the fauna is composed of animals of a very wide range of ecological origin. The first group includes the usual European forms, and even some warmth-loving forms such as the mountain lizard (Lacerta vivipera), viper (Vipera berus) and frog (Rana temporaria). The marine birds on the west coast of Norway are similar with respect to species composition and racial differentiation to those of southern Norway or the British Isles: razorbill, cormorant, shag, white-tailed eagle, black-backed gull and herring gull are even found far north of the Arctic Circle (nothing is known about the zone at the threshold of the true Arctic, where black-backed gull and herring gull meet yellow-footed gulls from the polar seas, and guillemot and Brünnich's guillemot meet). Other central European elements encountered are the starling, magpie, oyster catcher and the marine nematoceran, Clunio. These elements are mixed with forms that are usually designated "arctic", although they are also met much farther south in the mountains, and even in the Alps. Examples are provided by rock and willow ptarmigan, redpoll and ring ouzel. Snowy owls, long-tailed ducks, common scoter, whimbrel, black-throated diver, horned grebe, red-necked phalarope and arctic skua are found as far south as southern Norway (Hardangervidda, Dovre-Fjell). In addition, however, there are typical arctic forms that either do not occur south of the Arctic Circle or, if so, only in small numbers and far to the east, which is a type of distribution known to botanists as "northern unicentric". It is chiefly exhibit by a number of butterflies, although the northern redbacked vole, Clethrionimus rutilus, provides us with an example among the mammals, and it is also seen in birds such as the lesser white-fronted goose, the bar-tailed godwit (Limosa lapponica), the little stint and, to a lesser extent, the bean goose (Fig. 121). Only some of these forms turn up in the Norwegian part of our section, and most of them are first seen much farther east. Still farther to the east they are joined by animals typical of the Siberian tundra, like the red-throated pipit and the spotted redshank. The typical plant representatives of the northern unicentric type of distribution are Pedicularis hirsuta and Cassiope tetragona.

Towards the east, typical taiga organisms like the flying squirrel (Glaucomys), hawk owl, great grey owl, three-toed woodpecker, crossbill, nutcracker and Siberian jay appear in gradually increasing numbers whilst central European forms like the magpie, sparrow and starling gradually disappear.

The high summer temperatures favour the existence of a diverse collection of herbivorous insects, which show a tendency to explosive multiplication with devastating consequences for the birch forests. Operophtera brumata and Oporinia autumnata have been responsible for destroying large expanses of such forest, with the result that either the timberline has receded considerably (Fig. 97b, in the vicinity of Kevo) or the forest has taken many years to recover (near Abisko). Both species are also fairly numerous in localities far to the south, but they do not

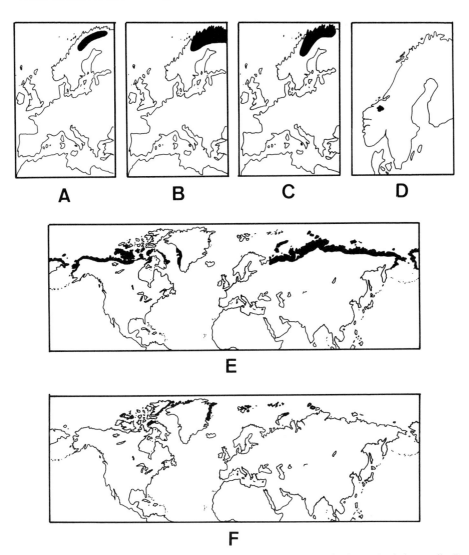

Fig. 121A-F. Distribution of arctic types of terrestrial plants and animals. **A** Agriades aquilo **B** Euphydryas iduna Dalm **C** Colias hecla lef **D** Artemisia norvegica Fr. **E** Stercorarius pomarinus **F** Pagophila eburnea. The range of distribution of these species nowhere extends south of the Arctic Circle (so-called northern unicentric species). Only (**D**) is a southern unicentric type of species

multiply in the same catastrophic manner. It seems that outbreaks of such dimensions are only possible under the special set of conditions prevailing on the northern limits of the birch forest. The summer temperatures are still high enough for the mass increases but the fauna is no longer of sufficient diversity to exert a buffering action against the vast numbers of moths. No mass outbreaks of inchworm have been reported from Tromsö, so that a reinvestigation at this point would probably bring interesting results. Anthills (Formica) in sufficient density suppress outbreaks of Oporinia (Laine and Niemelä, 1980).

The birch forest around Tromsö seems to be particularly abundant in large carabids, bumble-bees, honey-bees and wasps and it would therefore be of interest to make a comparison with birch forests farther inland. However, since no detailed studies are at present available, we shall have to confine our attention to the birch forests around Abisko (farther south) which already show considerable continental influence, and to those around Kevo. Large numbers of moisture-loving herbaceous plants grow beneath and between the trees of the wet birch forests of western Norway, in contrast to the various Ericaceae on the floor of the dry birch forests of Sweden and particularly of Finland. Tables 10, 11, 12, 13, and 14 give a survey of the fauna and flora. They reveal very clearly the typical pauperization of arctic regions with respect to large decomposers, and at the same time illustrate the changes taking place along our section from west to east.

The differences can only be explained by the fact that the part of western Norway north of the Arctic Circle receives less rain but more sunshine than the rest of western Norway. Moisture is available in adequate quantities, and water is never a limiting factor as it is at times in the tundra of north-east Finland. For this reason, as pointed out by Väisänen and Järvinen, (1977), the bird life in the west of our section is generally richer than in the east.

Unfortunately, the birds have not been studied with such thoroughness in the west as in the east, and we shall therefore have to take the summary given by Väisänen and Järvinen as being characteristic of the bird communities in the border regions between tundra and taiga in north Finland (Fig. 122). The most remarkable thing about the data is the very uneven decrease in diversity from south to north (from taiga to tundra). The greatest diversity is seen in alpine meadows with streams and ponds (see also Table 8).

Due to the briefness of the summer, birds that usually have a very long moulting season are forced to concentrate this into a few weeks and thus, in contrast to the situation in their more southern breeding localities, they may be entirely or partially unable to fly for a certain time. This happens to the bluethroat, willow warbler, and yellow and white wagtails in Finnish Lapland (Haukioja, 1971).

Only a few hundred kilometres to the south, in the immediate vicinity of the Arctic Circle, the fauna becomes still more diverse. From Müller's investigations in Messaure (particularly in 1970 and 1974; Müller, 1974) we are in a position to form a comprehensive picture of what is an incomparably rich fauna for arctic regions. A few examples are shown in Tables 13–15.

But even Hardangervidda, one of the best known "tundra" regions in southern Norway, and far to the south of the Arctic Circle, is extremely rich as compared with the northerly arctic tundra regions, and the fauna includes a wealth of southern forms such as snails, isopods, diplopods, deer, beavers, foxes and blackbirds. It is the possibility of strong sunshine during the summer that provides the conditions neccessary for supporting this comparatively rich fauna (Christiansen, 1961; Müller, 1970, 1974; Anonym, 1974; Wielgolaski, 1975).

Thus, in northern Scandinavia, extraordinarily varied arctic ecosystems are encountered within a very small space. Despite the many current investigations, real progress has so far been limited to small sections of the overall problem. The role played by animals is considerably more complex than that of plants and has not yet been thoroughly elucidated.

Table 10. Vertebrate species of the Torneträsk region. H – herbivore; H_g – granivore (seeds); H_h – grazer, browser, etc.; C – carnivore; C_i – "insectivore" (invertebrate feeder); C_v – predator; O – omnivore. A – abundant; C – common; U – uncommon; R – rare. (Johnson in Wielgolaski, 1975)

Ecosystem Component	Function	Relative abundance	Ecosystem Component	Function	Relative abundance
Pisces			Gavia arctic	C_v/C_i	U
Salmo trutta	C_i	C^c	Anas platyrhynchos	H, (O)	U
Salvelinus alpinus	C_i	A^c	A. crecca	H, (O)	C
Coregonus pidschian	C_i/C_v	A^c	Aythya marila	C_i	C
Thymallus thymallus	C_i/C_v	C	Melanitta fusca	C_i	C
Esox lucius	C_v	U	M. nibra	C_i	C
Phoxinus phoxinus	C_i/C_v	C	Mergus merganser	C_v	U
Gasterosteus pungitius	C_i	C	Aquila chrysaëtos	C_v	R
Lota lota	C_v/C_i	C	Buteo lagopus	C_v	A^a
Perca fluviatilis	C_i/C_v	C	Falco peregrinus	C_v	R
Cottus poecilopus	C_i/C_v	U	F. rusticolus	C_v	R
Amphibia			F. columbarius	C_v	U^a
Rana t. temporaria	C_i	C	F. tinnunculus	C_v	C^a
Reptilia			Lagopus lagopus	H/C_i	A^a
Lacerta vivipara	C_i	U	L. mutus	H/C_i	A^a
Aves			Grus grus	O	U
Gavia a. arctica	C_v	U^a	Vanellus vanellus	C_i	U
Gavia stellata	C_v	C^a	Charadrius hiaticula	C_i	C
Podiceps a. auritus	C_v/C_i	U^a	Pluvialis apricaria	C_i	A
Anas p. platyrhynchos	H_g/H_h	U^a	Eudromias morinellus	C_i, (O)	A
Anas c. crecca	H/C_i	A^a	Gallinago gallinago	C_i, (O)	U
Anas penelope	H_h	C^a	G. media	C_i, (O)	R
Anas acuta	H_h	U^a	Numenius phaeopus	O	R
Authya m. marila	C_i	U^a	Tringa glareola	C_i	C
Aythya fuligula	C_i	C^a	T. totanus	C_i	C
Clangula hyemalis	C_i	U^a	Actitis hypoleucos	C_i	C
Melanitta f. fusca	C_i	C^a	Calidris maritima	C_i	C
Melanitta n. nigra	C_i/H	C	C. temminckii	C_i	C
Bucephala c. clangula	C_i	U	C. alpina	C_i	C
Mergus s. serrator	C_v	C^a	Limicola falcinellus	C_i	R
Mergus m. merganser	C_v	R	Philomachus pugnax	C_i, (O)	U
Anser erythropus	H_h	R	Phalaropus lobatus	C_i	R
Anser f. fabalis	H_h	R	Stercorarius longicaudus	O, (C_v)	R
Cygnus cygnus	H_h	R	Larus canus	O	C
Aquila c. chrysaëtos	C_v	R	Cuculus canorus	C_i/O	C
Buteo l. lagopus	C_v	C^a	Nyctea scandiaca	C_v	R^a
Accipiter n. nisus	C_v	R^b	Asio flammeus	C_v	C^a
Haliaeetus albicilla	C_v	R	Alauda arvensis	O	R
Pandion h. haliaëtus	C_v	R	Eremophila alpestris	C_i, (O)	C
Falco p. peregrinus	C_v	C	Delichon urbica	C_i	U^b
Falco r. rusticolus	C_v	R^a	Corvus corax	O	A
Falco columbarius aesalon	C_v	C	Vanellus vanellus	C_i	R^b
Falco t. tinnunculus	C_v/C_i	C^a	Charadrius hiaticula tundrae	C_i	C^a
Lagopus l. lagopus	H_g/H_h	$U-A^a$	Eudromias morinellus	C_i	C
Lagopus m. mutus	H_g/H_h	R–C	Pluvialis apricaria altifrons	C_i/H	C^a
Lyrurus t. tetrix	H_g/H_h	R	Gallinago g. gallinago	C_i	C^a
Tetrao u. urogallus	H_h	R	Gallinago media	C_i	R^a
Grus g. grus	O	U			

Table 10 (continued)

Ecosystem Component	Function	Relative abundance	Ecosystem Component	Function	Relative abundance
Lymnocryptes minimus	C_i	R^a	Parus major	O	R^b
Scolopax rusticola	C_i	R	Cinclus cinclus	C_i, (O)	U
Numenius a. arquata	C_i/H_g	$R^{a,b}$	Turdus pilaris	O	C
Numenius p. phaeopus	C_i/H_g	C^a	T. philomelos	O	R
Tringa ochropus	C_i	$R^{a,b}$	T. iliacus	O	R
Tringa glareola	C_i	C^a	T. torquatus	O	U
Tringa hypoleucos	C_i	A^a	T. merula	O	R
Tringa t. totanus	C_i	C^a	Oenanthe oenanthe	C_i	A
Tringa erythropus	C_i	R	Luscinia suecica	C_i, (O)	U
Tringa nebularia	C_i	C^a	Phylloscopus trochilus	C_i	U
Calidris maritima	C_i	R	Anthus pratensis	C_i	A
Calidris temminckii	C_i	C^a	Motacilla alba	C_i	C
Calidris a. alpina	C_i	R	Sturnus vulgaris	O	C^b
Philomachus pugnax	C_i/H_g	U^a	Carduelis flavirostris	H, (O)	C
Phalaropus lobatus	C_i	A^a	C. flammea	H, (O)	U
Stercorarius longicaudus	C_v	$R–C^a$	Emberiza schoeniclus	C_i, (O)	U
Larus f. fuscus	C_v/C_i	R^a	Calcarius lapponicus	O	C
Larus a. argentatus	C	U^a	Plectrophenax nivalis	H, (O)	A
Larus c. canus	C	C^a	Passer domesticus	O	R^b
Larus ridibundus	C_i	U	Mammalia		
Sterna paradisaea	C_i	R^a	Sorex araneus	C_i	C
Cuculus canorus	C_i	C^a	S. minutus	C_i	U
Bubo b. bubo	C_v	R	Neomys fodiens	C_i	U
Nycted scandiaca	C_v	R–C	Lepus timidus	H_h	A
Surnia u. ulula	C_v	R–C	Castor fiber	H_h	U
Asio o. otus	C_v	R^b	Lemmus lemmus	H, (O)	A^a
Asio f. flammeus	C_v	R–C	Clethrionomys glareolus	H	C^a
Aegolius f. funerus	C_v	R^b	C. rufocanus	H	A^a
Apus a. apus	C_i	R^a	Microtus agrestis	H	U^a
Dendrocopos m. minor	C_i	U	M. oeconomus	H	A^a
Picoidest t. tridactylus	C_i	R	Arvicola terrestris	H	R
Dryocopus m. martius	C_i	R^b	Vulpes vulpes	O	A
Alauda a. arvensis	H_g/C_i	R^a	Alopex lagopus	O	$U^{a,c}$
Eremophila alpestris flava	H_g/C_i	U	Mustela vison	C_v	
Delichon u. urbica	C_i	C^c	M. erminea	C_v	C^a
Riparia r. riparia	C_i	R^a	M. nivalis	C_v	$C^{a,d}$
Corvus c. corax	O	$C^{a,c}$	Gulo gulo	C_v	R^e
Corvus corone cornix	O	$C^{a,c}$	Lynx lynx	C_v	R^e
Pica pica fennorum	O	$C^{a,c}$	Alces alces	H_h	U^e
Perisoreus i. infaustus	O	R	Rangifer tarandus	H_h	A
Pica pica	O	R^b	Capreolus capreolus	H_h	R^e

[a] Cyclic
[b] Partly synantropic
[c] Population in slight increase
[d] M. nivalis or M. rixosa, systematic position not clear
[e] Probably not breeding within the mountain plateau

Table 11. Vertebrate species on the three IBP sites at Kevo, usually breeding within the area. H_g – granivore (seeds), H_h – grazer, browser, H – herbivores, C_i – carnivores on invertebrates, V_v – carnivores on vertebrates, O – omnivores: A – abundant, C – common, U – uncommon, R – rare. (Haukioja in Wielgolaski, 1975)

Ecosystem Component	Function	Relative abundance		
		Pine forest	Birch forest	Heath
Amphibia				
Rana temporaria	C_i	R	C	
Reptilia				
Lacerta vivipara	C_i	U	R	
Aves				
Aquila chrysaëtos	C_v	R	R	
Buteo lagopus	C_v	U	C	C
Accipiter gentilis	C_v	R		
Pandion haliaëtus	C_v	U	U	
Falco rusticolus	C_v			R
Falco columbarius	C_v	C	U	R(?)
Falco tinnunculus	C_v	U		
Lagopus lagopus	H_g/H	R	C	U
Lagopus mutus	H_g/H			U
Tetrao urogallus	H_g/H	U		
Pluvialis apricaria	C_i			C
Eudromias morinellus	C_i			C
Numenius phaeopus	C_i		R(?)	U
Stercorarius longicaudus	C_v			C
Cuculus canorus	C_i		U	
Surnia ulula	C_v	R(?)	U	
Dendrocopos major	C_i	R		
Dendrocopos minor	C_i		R	
Picoides tridactylus	C_i		R	
Corvus corax	O	C	C	U
Perisoreus infaustus	O	C	U(?)	
Parus major	C_i/O	C	C	
Parus cinctus	C_i/O	C	C	
Parus montanus	C_i/O	C	C	
Turdus pilaris	O	C	C	
Turdus philomelos	O	R(?)	R	
Turdus iliacus	O	C(?)	A	
Oenanthe oenanthe	C_i			A
Phoenicurus phoenicurus	C_i	C	C	
Luscinia svecica	C_i		A	
Erithacus rubecula	C_i		R	
Phylloscopus trochilus	C_i	C(?)	A	
Phylloscopus borealis	C_i		R	
Muscicapa striata	C_i	C(?)	C	
Ficedula hypoleuca	C_i	C(?)	C	
Prunella modularis	C_i/H_g		C	
Anthus pratensis	C_i		A	A
Anthus trivialis	C_i	C(?)	U	
Anthus cervinus	C_i			U
Motacilla alba	C_i		C	C
Motacilla flava	C_i		C	C
Carduelis flammea	H_g/C_i	C(?)	A	
Pyrrhula pyrrhula	H_g	R		

Table 11 (continued)

Ecosystem Component	Function	Relative abundance		
		Pine forest	Birch forest	Heath
Pinicola enucleator	O	R	C	
Fringilla montifringilla	C_i/H_g	U	A	
Emberiza schoeniculus	C_i		C	
Calcarius lapponicus	C_i/H_g			A
Plecrophenax nivalis	O			C
Mammalia				
Sorex araneous	C_i	U(?)	C	U(?)
Sorex minutes	C_i	C	C	U(?)
Lepus timidus	H_h	C	C	
Sciurus vulgaris	H_g/H	R–A		
Lemmus lemmus	H		R–C	R–A
Clethrionomys rutilus	H	R–C(?)	R–C	R–C
Clethrionomys rufocanus	H		R–C	R–C
Microtus oeconomus	H	C	R–C	R–C
Vulpes vulpes	C_v		C	U(?)
Alopex lagopus	C_v	C	R	R
Mustella erminea	C_v	C(?)	C	
Mustella rixosa	C_v	R	C	
Gulo gulo	C_v	U	R	R
Alces alces	H	C	U	
Rangifer tarandus	H_h		A	A

Table 12. Density of ground layer mesofauna (number per m^2 with S.E.) near Kevo, 1973. (Koponen and Ojala in Wielgolaski, 1975)

	June		July		August	
	Pine forest	Birch forest	Pine forest	Birch forest	Pine forest	Birch forest
Araneae	112.0± 47.3	267.2± 67.9	106.0± 32.3	228.0±36.9	120.0±11.7	220.7±23.1
Linyphiidae and Theridiidae	98.7± 48.1	238.7± 64.3	90.0± 31.1	200.0±39.3	84.8±11.9	173.3±21.2
Phalangida	0	12.0± 9.3	0	1.0± 1.0	0	0.7± 0.7
Insecta adults	704.0±242.8	381.3± 42.0	498.8± 61.2	315.0±28.1	176.0±32.4	180.7±21.0
Coleoptera ad.	24.0± 12.2	53.0± 12.6	34.0± 10.2	59.0± 9.7	36.8± 7.9	68.7±11.3
Diptera ad.	120.0± 29.1	189.3± 32.8	94.0± 24.3	78.0±17.5	11.2± 4.2	10.0± 4.8
Hemiptera	496.0±212.4	118.7± 35.7	142.0± 34.8	109.0±18.7	43.2± 8.6	66.7±11.9
Insecta larvae	376.0±116.6	314.7± 43.1	316.5± 38.0	215.0±42.7	409.6±77.3	432.0±70.9
Coleoptera l.	80.0± 17.0	213.3± 30.3	155.4± 23.2	138.0±33.5	198.4±51.5	134.7±16.4
Diptera larv.	296.0±117.4	101.3± 23.9	150.0± 42.3	73.0±17.4	185.6±64.0	286.0±66.2
Total mesofauna	1197.3±340.1	975.7±124.8	923.3±104.5	760.0±67.2	710.6±80.9	839.9±70.2
n	6	12	8	16	10	24

Table 13. Butterflies and moths of the Messaure region (Sweden). (Müller, 1974)

Papilionidae
Papilio machaon

Pieridae
Pieris napi
Anthocaris cardamines
Colias palaeno
Leptidea sinapis

Nymphalididae
Aglais urticae
Nymphalis antiopa
Proclossiana eunomia
Clossiana selene
C. euphrosyne
Boloria aquilonaris

Satyridae
Erebia embla
Oeneis norna
Coenonympha pamphilus
C. tullia
Pararge petropolitana

Lycaenidae
Callophrys rubi
Palaeochrysophanus hippothoe
Lycaena helle
Celastrina argiolus
Vacciniina optilete
Lycaeides idas
Cyaniris semiargus
Lysandra amanda
Polyommatus icarus

Hesperiidae
Carterocephalus palaemon

Sphingidae
Amorpha populi
Celerio galii

Notodontidae
Dicranura vinula
Notodonta ziczac
N. dromedarius
Odontosia sieversi
Pterostoma palpinum
Pygaera pigra

Lasiocampidae
Trichiura crataegi
Poecilocampa populi
Epicnaptera ilicifolia

Lymantriidae
Orgyia antiqua
Dasychira fascelina
Stilpnotia salicis

Drepanidae
Drepana lacertinaria
D. falcataria

Polyplocidae
Palimpsestis duplaris
Polyploca cinerea

Noctuidae
Acronycta leporina
A. auricoma
Euxoa recussa
Rhyacia mendica
R. dahlii
R. rubi
R. alpicola
R. cuprea
R. augur
Pachnobia tecta
Aplectoides speciosa
Anomogyna laetabilis
A. rhaetica
Eurois occulta
Cerastis sobrina
Polia dissimilis
P. glauca
Harmodia rivularis
Cerapteryx graminis
Monima gothica
Sideridis conigera
S. impura
Brachionycha nubeculosa
Cloantha solidaginis
Bombycia viminalis
Hillia iris
Litophane lambda
Xylina vetusta
Crypsedra gemmea
Amathes suspecta
Cosmia lutea
C. icteritia
Amphipyra tragopogonis
Parastichlis rurea
P. lateritia
Procus haworthii
Crymodes maillardi
Lithomoia rectilinea
Elaphria morpheus
Athetis palustris
Apamea fucosa
Hydroecia micacea
Enargia paleacea
Arenostola pygmina
Anarta cordigera
Gonospileia glyphica
Syngrapha interrogationis

Plusia putnami
P. macrogamma
P. confusa
Scoliopteryx libatrix
Herminia tentacularia
Hypena proboscidalis

Geometridae
Brephos partenias
Epirrhantis diversata
Geometra papillionaria
Jodis putata
Cosymbia albipunctata
Scopula ternata
S. immorata
Sterrha serpentata
Ortholitha chenopodiata
Carsia sororiata
Nothopteryx carpinata
Operophthera brumata
Oporinia autumnata
Lygris prunata
L. testata
L. populata
Cidaria bicolorata
C. obeliscata
C. juniperata
C. truncata
C. citrata
C. munitata
C. fluctuata
C. annotinata
C. montanata
C. spadicearia
C. ferrugata
C. caesiata
C. luctuata
C. hastata
C. subhastata
C. hastulata
C. albulata
C. ruberata
C. coerulata
C. furcata
Pelurga comitata
Eupithecia bilunulata
E. icterata
E. succenturiata
E. satyrata
E. intricata
E. vulgata
E. virgaureata
E. conterminata
E. indigata
E. nanata
E. sobrinata

Table 13 (continued)

Coenocalpe lapidata	Semiothisa liturata	**Arctiidae**
Arichanna melanaria	S. wauaria	Comacla senex
Lomaspilis marginata	Itame fulvaria	Philea irrorella
Cabera pusaria	I. loricaria	Phragmatobia fuliginosa
C. exanthemata	Chiasma clathrata	Parasemia plantaginis
Ellopia fasciaria	Lycia hirtaria	Diacrisia sannio
Selenia bilunaria	Gnophos sordaria	Arctia caja
Epione repandaria	Isturga carbonaria	**Hepialidae**
Hypoxystis pluviaria	Ematurga atomaria	Hepialus fusconebulosus

Table 14. Caddis flies of the Messaure region (Sweden). (Göthberg in Müller, 1974)

Rhyacophilidae
Rhyacophila fasciata Hagen
R. nubila (Zetterstedt)
R. obliterata McLachlan
A. lapponicus (Zetterstedt)
A. thedenii (Wallengren)
Arctopora trimaculata (Zetterstedt)
Lenarchus productus (Morton)
Rhadicoleptus alpestris (Kolenati)
Potamophylax cingulatus (Stephens)
P. latipennis (Curtis)
P. nigricornis (Pictet)
Halesus digitatus (Schrank)
H. radiatus interpunctatus (Zetterstedt)
H. tesselatus (Rambur)
Micropterna lateralis (Stephens)
M. sequax McLachlan
Hydatophylax infumatus (McLachlan)
Chaetopteryx sahlbergi McLachlan
Ch. villosa (Fabricius)
Annitella obscurata (McLachlan)
Chilostigma sieboldi McLachlan
Brachypsyche sibirica (Martynov)

Goeridae
Silo pallipes (Fabricius)

Lepidostomatidae
Lepidostoma hirtum (Fabricius)

Leptoceridae
Arthripsodes alboguttatus (Hagen)
A. annulicornis (Stephens)
A. aterrimus (Stephens)
A. cinereus (Curtis)
A. commutatus (Rostock)
A. dissimilis (Stephens)
A. excisus (Morton)
A. fulvus (Rambur)
A. nigronervosus (Retzius)
A. perplexus (McLachlan)
Mystacides azurea (Linnaeus)

M. longicornis (Linnaeus)
M. nigra (Linnaeus)
Triaenodes bicolor (Curtis)
T. forsslundi Tjeder
Oecetis lacustris (Pictet)
O. ochracea (Curtis)
O. testacea (Curtis)

Sericostomatidae
Sericostoma personatum (Spence)

Beraeidae
Beraeodes minutus (Linnaeus)

Molannidae
Molanna albicans (Zetterstedt)
M. angustata Curtis
M. carbonaria McLachlan
M. submarginalis McLachlan
Molannodes tincta (Zetterstedt)

Glossosomatidae
Synafophora intermedia (Klapálek)
S. nylanderi (McLachlan)
Agapetus ochripes Curtis

Hydroptilidae
Ithytrichia lamellaris Eaton
Oxyethira distinctella McLachlan
O. ecornutta Morton
O. flavicorns (Pictet)
O. friti (Klapálek)
O. tristella Klapálek
Oxytrichia mirabilis (Morton)
Hydroptila forcipata (Eaton)
H. pulchricornis Pictet
H. simulans Mosely
H. tineoides Dalman
H. vectis Curtis
Agraylea cognatella McLachlan

Philopotamidae
Philopotamus montanus (Donovan)
Wormaldia subnigra McLachlan

Table 14 (continued)

Arctopsychidae	**Limnephilidae**
Arctopsyche ladogensis Kolenati	Apatania auricula (Forsslund)
	A. hispida (Forsslund)
Hydropsychidae	A. muliebris McLachlan
Hydropsyche nevae Kolenati	A. stigmatella (Zetterstedt)
H. ornatula McLachlan	A. wallengreni McLachlan
H. pellucidula (Curtis)	A. zonella (Zetterstedt)
H. saxonica McLachlan	Ironoquia dubia (Stephens)
H. silfvenii Ulmer	Limnephilus algosus (McLachlan)
H. siltalai Döhler	L. auricula Curtis
Cheumatopsyche lepida (Pictet)	L. binotatus Curtis
	L. borealis (Zetterstedt)
Polycentropodidae	L. centralis Curtis
Neureclipsis bimaculata (Linnaeus)	L. coenosus Curtis
Plectrocnemia conjuncta Martynov	L. decipiens (Kolenati)
P. conspersa (Curtis)	L. diphyes McLachlan
P. flavomaculatus (Pictet)	L. dispar McLachlan
P. irroratus (Curtis)	L. elegans Curtis
Holocentropus dubius (Rambur)	L. externus Hagen
H. insignis Martynov	L. extricatus McLachlan
H. picicornis (Stephens)	L. femoratus (Zetterstedt)
Cyrnus flavidus McLachlan	L. fenestratus (Zetterstedt)
C. trimaculatus (Curtis)	L. flavicornis (Fabricius)
	L. fuscicornis (Rambur)
Psychomyiidae	L. germanus McLachlan
Tinodes waeneri (Linnaeus)	L. incisus Curtis
Psychomyia pusilla (Fabricius)	L. nebulosus Kirby
Lype phaeopa (Stephens)	L. nigriceps (Zetterstedt)
	L. pantodapus McLachlan
Phryganeodae	L. picturatus McLachlan
Agrypnia obsoleta (Hagen)	L. politus McLachlan
A. pagetana Curtis	L. quadratus Martynov
A. picta Kolenati	L. rhombicus (Linnaeus)
A. principalis (Martynov)	L. scalenus Wallengren
A. varia (Fabricius)	L. sericeus (Say)
Phryganea bipunctata (Retzius)	L. sparsus Curtis
Ph. grandis Linnaeus	L. stigma Curtis
Oligotricha lapponica (Hagen)	L. vittatus (Fabricius)
Hagenella clathrata (Kolenati)	Grammotaulus atomarius (Fabricius)
Semblis atrata (Gmelin)	G. signatipennis McLachlan
S. phalaenoides (Linnaeus)	Nemotaulius punctatolineatus (Retzius)
Oligostomis reticulata (Linnaeus)	Anabolia brevipennis (Curtis)
	A. concentrica (Zetterstedt)
Brachycentridae	A. soror McLachlan
Brachycentrus subnubilus Curtis	Asynarchus contumax McLachlan
Micrasema gelidum McLachlan	
M. setiferum (Pictet)	

Table 15. The fish of the Messaure region. (Müller, 1974)

Perca fluviatilis	Salmo trutta	Phoxinus phoxinus
Leuciscus rutilis	Salvelinus alpinus	Petromyzon planeri
Esox lucius	Salvelinus fontinalis	Thymallus thymallus
Lota lota	Cottus gobio	Pungitius pungitius
Coregonus larvaretus	Cottus poecilopus	Acerina cernua

The important question as to the time and place at which reindeer were domesticated in the Old World has still received no satisfactory answer. On the one hand, sources going back to the days of the Vikings, suggest that the old kings of Norway possessed reindeer herds, whereas other sources suggest that it was the Lapps who first began to domesticate reindeer about 250 years ago. This implies a possible discrepancy of at least 700 years in the length of time for which the Scandinavian tundra and tundra-like ecosystems have been exposed to the specific influence of the domesticated reindeer. Reindeer are beyond doubt the major and most conspicuous herbivores above the timberline in Scandinavia. Three types can be distinguished. The forest reindeer is very rare and lives alone and well hidden in the forests of Karelia and farther to the east, usually only in small numbers. It is also found in eastern Finland, again in very low numbers. It never seems to leave the forests voluntarily, does not mate with the Fjell reindeer that pour into its environment in winter, but behaves as a distinct species.

The Fjell reindeer of Scandinavia, the only form that concerns us here, are all domesticated and no wild form any longer exists on the Fjells. There are said to be about 2 million of these domesticated animals in the Soviet Union, and about 500,000 in Fennoscandia. They usually live a semi-wild state in herds of several hundred individuals. Since the majority of the males are castrated their social behaviour differs markedly from that of the wild forms. A reindeer requires about 450 kg plant biomass (dry weight) annually, whereby the same areas are seldom grazed in succession. This means that the slow-growing lichens of the genera Cetraria and Cladonia, that constitute the winter food of the reindeer, often have several years in which to regenerate before revisitation. These domesticated reindeer undertake characteristic migrations. They spend the summer on the high-lying treeless fjells and often seek the coolness of snowy areas. In winter they retire to lower situations, often penetrating far into the forests. The large herds frequently become disrupted in the process and the animals then live alone or in small groups like the forest reindeer, often consuming tree lichens. The distance of alarm is less than on the open fjell.

In southern Norway; the only wild European reindeer found nowadays are in the vicinity of Hardangervidda and Dovre-Fjell. On the basis of counts made from the air it is estimated that about 10,000 individuals make up the herd on Hardangervidda. These are no longer pure wild forms, however, but are the offspring of what were earlier in part domesticated reindeer. The majority of the animals live in groups of 100–500, and particularly in July tend to join up to form very large herds. Within the large herds there is a total integration of the sexes and of all age groups, whereas the smaller groups consist either of stags alone, or of females with their calves. The smaller groups are more common at times other than the height of summer. It is estimated that there are 1–2 animals per km^2, which probably corresponds to a grazing density of 2 animals per square kilometre. The effect of the grazing itself on the vegetation is relatively small, but about three times as many plants are destroyed by trampling. In all, the reindeer of Hardangervidda, at this density, probably consume and destroy no more than 0.6% of the primary production. For herbivores this is a low figure, and is probably connected with the fact that the animals do not retire to the lower-lying forests in winter, but remain on the open, lichen-covered high plateaus (Fig. 123).

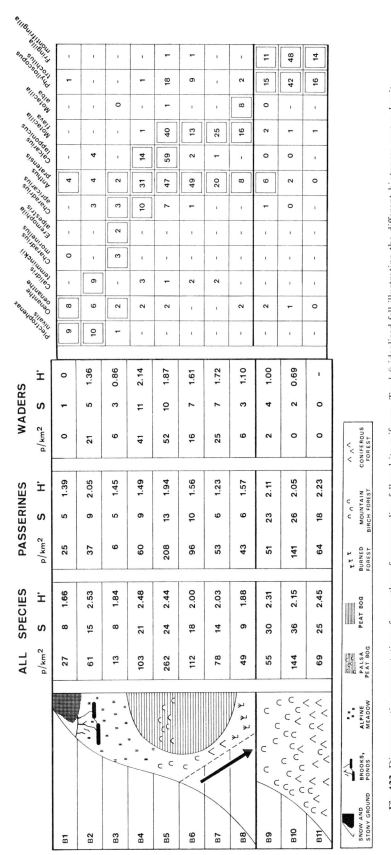

Fig. 122. Diagrammatic representation of a northern fennoscandian fell and its avifauna. *Top left* idealised fell illustrating the different biotopes; *center* density (nests/km^2) species richness (S) and diversity (H') of the avifauna; *right* 12 selected species and their breeding distribution in the fell. (Järvinen, 1976)

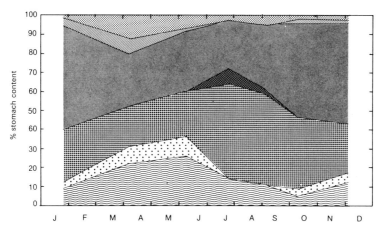

Fig. 123. Plants eaten by wild reindeer in southern Norway (Hardangervidda) in the course of the year. From top to bottom: Mosses, epilithic lichens, epigaeic lichens, forbs, graminids, litter, woody plants. (Gaare and Skogland in Wielgolaski, 1975)

Well-developed roads, railways and pipelines apparently represent an insuperable obstacle to all wild forms of reindeer. The main road connecting Oslo and Trondheim is seldom crossed by them, and wild caribous behaved in exactly the same way during the construction of the Alaska pipeline.

2. Arctic Alaska: Comparison Between Prudhoe Bay and Point Barrow

Although at about the same geographical latitude as the section through northern Scandinavia, the well known arctic stations of Prudhoe and Point Barrow in Alaska represent a completely different environment (Fig. 124). Their summers are colder, whereas the winters are roughly the same as in Kevo. They are pure, treeless tundra regions, with only occasional stands of grey willow. Taiga or even taiga-like patches of forest are several hundred kilometres farther to the south. Although both points are equidistant from the Arctic Circle and both are situated on the north coast of Alaska, Prudhoe and Point Barrow are very different. Prudhoe, where the sea ice scarcely breaks up at all in summer, has a much more continental type of climate than Barrow (Fig. 118): its winters are slightly colder and its summers slightly warmer. But even such small differences in temperature are responsible for very large differences in the invertebrate fauna. This is well covered in a monograph by Brown (1975). Prudhoe has a much richer fauna than Barrow, especially with respect to butterflies and Dolichopodids (Tables 16–18). Whether there are also differences in population density, which ought theoretically to be the case, is not yet certain, although the data so far available suggest that this may be so (Fig. 125; Table 16). Differences in the homeothermic fauna are not so large. Lemmings play a major role in both places. In Barrow they exhibit conspicuous regular four-year

Table 16a, b. The population density and breeding success of a number of bird species at Barrow and Prudhoe **a**, and a list of bird species found in the Prudhoe region **b**

Species	Prudhoe Bay		Barrow	
	Density (nests km^{-2})	Success	Density (nests km^{-2})	Success
C. alpina	4.6	0.5	13.9	0.72
C. pusilla	39.8	0.68	9.8	0.73
C. bairdii	—	—	24.8	0.39
C. melanotos	5.7	0.67	13.6	0.69[a]
T. subruficollis	4.3	0.50	—	—
P. fulicarius	29.7	0.57	[26.4	0.50][b]
L. lobatus	5.0	0.67	—	—
C. lapponicus	7.7	0.60	[30.0	0.63][c]
P. nivalis	—	—	[15.0	0.80][d]
Overall	96.7	0.62	[133.5	0.65]

[a] Calidris species data based on Norton (1973)
[b] Phalarope data from 1971 at Barrow only–preliminary
[c] Longspur information from T.W. Custer (pers. comm.)
[d] Snow bunting data from 1971 IBP census plot only–incomplete

Table 16b

Gavia adamsii	Yellow-billed loon	C, SB
G. arctica pacifica	Arctic loon	KB
G. stellata	Red-throated loon	KB
Cygnus columbianus	Whistling swan	KB
Branta canadensis minima	Canada goose	KB
Branta nigricans	Black brant	KB
Anser albifrons frontalis	White-fronted goose	KB
Chen caerulescens caerulescens	Snow goose	KB
Anas platyrhynchos platyrhynchos	Mallard	C
Anas acuta	Pintail	KB
Anas americana	American wigeon	C
Anas clypeata	Shoveler	C
Aythya marila nearctica	Greater scaup	C
Clangula hyemalis	Oldsquaw	KB
Polysticta stelleri	Steller's eider	SB
Somateria mollissima v-nigra	Common eider	KB
Somateria spectabilis	King eider	KB
Somateria fischeri	Spectacled eider	KB
Melanitta perspicillata	Surf scoter	C
Mergus serrator serrator	Red-breasted merganser	C
Buteo lagopus sanctijohannis	Rough-legged hawk	C
Lagopus lagopus alascensis	Willow ptarmigan	KB
L. mutus nelsoni	Rock ptarmigan	KB
Grus canadensis canadensis	Sandhill crane	M
Charadrius semipalmatus	Semipalmated plover	SB
Pluvialis dominica dominica	American golden plover	M, SB
P. squatarola	Black-bellied plover	KB

Table 16b (continued)

Arenaria interpres (subsp)	Ruddy turnstone	M, SB
Capella gallinago delicata	Common snipe	C
Micropalama himantopus	Stilt sandpiper	M
Limnodromus scolopaceus	Long-billed dowitcher	M
Calidris alpina sakhalina	Dunlin	KB
C. pusilla	Semipalmated sandpiper	KB
C. bairdii	Baird's sandpiper	M, SB
C. mauri	Western sandpiper	C
C. melanotos	Pectoral sandpiper	KB
C. alba	Sanderling	M
Tryngites subruficollis	Buff-breasted sandpiper	KB
Phalaropus fulicarius	Red phalarope	KB
Lobipes lobatus	Northern phalarope	KB
Stercorarius pomarinus	Pomarine jaeger	SB
S. parasiticus	Parasitic jaeger	SB
S. longicaudus	Long-tailed jaeger	M
Larus hyperboreus barrovianus	Glaucous gull	KB
Xema sabini sabini	Sabine's gull	SB
Sterna paradisea	Arctic tern	SB
Nyctea scandiaca	Snowy owl	SB
Asio flammeus flammeus	Short-eared owl	SB
Corvus corax principalis	Common raven	M
Motacilla flava tschutschensis	Yellow wagtail	M
Acanthis (sp.)	Redpoll	SB
Calcarius lapponicus alascensis	Lapland longspur	KB
Plectrophenax nivalis nivalis	Snow bunting	KB

KB Known breeding in Prudhoe region
SB Suspected breeding
M Regular movement or migration through region
C Casual movement or migration through region

Table 17. The abundance of the more important soil organisms at Barrow and Prudhoe. (Numbers per m^2)

Group	Barrow			Prudhoe Bay			
	Maximum	Minimum	Mean	Plot 1	Plot 6	Plot 7	Plots 4/5
Acarina							
Prostigmata	42,500	4,350	18,000	63,200	1,090	38,700	
Mesostigmata	7,080	603	3,260	1,090	109	3,710	
Cryptostigmata	38,900	1,600	20,000	11,900	5,790	4,040	
Total Acarina			41,300	76,200	7,000	46,500	
Collembola							
Entomobryidae	172,000	22,400	83,000	11,900	2,180	31,100	
Poduridae	33,200	1,150	8,110	3,820	436	21,400	
Sminthuridae	2,850	0	1,170	3,490	0	7,750	
Total Collembola			92,300	19,200	2,620	60,300	
Enchytraeidae	95,300	11,600	46,900	15,900	30,900	20,900	32,900

Table 18. A comparison of the insect fauna of Barrow and Prudhoe. (Brown, 1975)

Barrow	Prudhoe Bay
Diptera: Tipulidae	(Det. by G.C. Byers and C.P. Alexander)
Tipula carinifrons Holm.	Tipula begrothiana Alex.
Tipula aleutica Alex.	Tipula arctica Curtis
(Det. by C.P. Alexander)	Tipula pribilofensis Alex.
	Tipula diflava Alex.
	Tipula besselsi Osten Sacken
	Tipula macleani (sp. nov.) Alex.
	Prionocera parii Osten Sacken
Prionocera gracilistyla Alex.	Prionocera gracilistyla Alex.
	Nephrotoma lundbecki (Nielsen)
Pedicia hannai antennata Alex.	Pedicia hannai antennata Alex.
	Erioptera kluane Alex.
	Erioptera forcipata Lundstrom
	Limnophila sp. nov.
Diptera: Culicidae	(Det. by R. Gorham)
Aedes nigripes	Aedes cataphylla
	Aedes impiger
	Aedes nigripes
Diptera: Dolichopodidae	(Det. by F. Harmston)
	Dolichopus amnicola
	Dolichopus consanguineus
	Dolichopus obcordatus
	Dolichopus eudactylus
	Dolichopus ramifer
	Dolichopus plumipes
	Dolichopus occidentalis
	Dolichopus aldrichii
	Dolichopus humilis
	Campsicnemus nigripes
	Hydrophorus gratiosus
	Hydrophorus sodalis
	Hydrophorus signiterus
Hydrophorus fumipennis	Hydrophorus fumipennis
	Gymnopternus californicus
	Aphrosylus nigripennis
	Aphrosylus fumipennis
	Aphrosylus praedator
	Raphium tripartitum
	Raphium sp.
Lepidoptera: Pieridae	(Det. by K.W. Philip)
Colias palaeno	
	Colias hecla
	Colias thula
	Colias nastes
Lepidoptera: Papilionidae	
Papilio machaon	
(Det. by K.W. Philip)	
Lepidoptera: Lycaenidae	(Det. by K.W. Philip)
	Lycaeides argyrognomon
	Agriades aquilo
Lepidoptera: Nymphalidae	
(Det. by K.W. Philip)	(Det. by K.W. Philip)
Boloria frigga	Boloria frigga
Boloria polaris	Boloria polaris
Boloria chariclea	Boloria chariclea
	Boloria napaea
Lepidoptera: Satyridae	(Det. by K.W. Philip)
	Oenis melissa
	Erebia rossii
	Erebia fasciata

Fig. 124. Tundra types of North America. **a** Prudhoe, with caribous (phot. Dr. D. R. Klein).
b Vegetation near Inuvik (phot. Dr. L. Bliss). **c** Devon Island, with muskoxen (phot. Dr. L. Bliss).
d Ellesmere Island, with Arctic hare (phot. Dr. D. Parmelee). **e** Ellesmere Island, with muskoxen (phot. Dr. D. Parmelee)

cycles, which have been exhaustively studied, whereas in Prudhoe they do not. The arctic lemming is the more common in Prudhoe whereas the brown form dominates in Barrow. The fact that in Prudhoe the lemming populations never achieve the numbers seen every four years in Barrow can scarcely be put down to differences in pressure from predators, since the same species are found in both localities, i.e., short-eared owl, snowy owl, long-tailed skua, weasel, arctic fox, wolf and grisly bear. Citellus undulatus, the ground squirrel, also occurs in both localities, but muskoxen are not found in Alaska. Caribou must have been consistently hunted in Point Barrow in earlier times and are now practically absent, whereas they are seen in Prudhoe. There does appear to be a clear connection between the occurrence of reindeer and the bloodsucking of mosquitoes, which apparently can only feed if temperature and wind speed are favourable (Fig. 126). But under such conditions the reindeer usually withdraw from threatened areas, and in Prudhoe there is a strong tendency for the reindeer to migrate towards the coast under the pressure of the mosquitoes (stronger winds and lower temperatures than inland), and to return

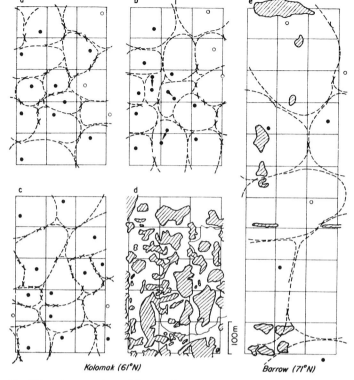

Fig. 125a-e. A comparison of the population density of the dunlin in a favourable (Kolomak river) and less favourable (Barrow) region of Alaska. (Holmes in Watson, 1970)

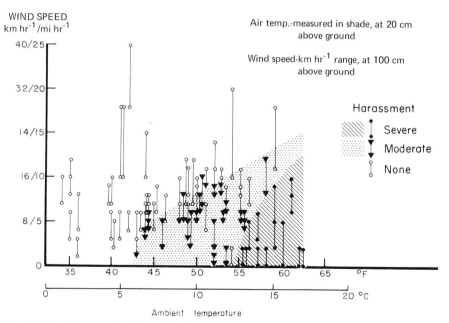

Fig. 126. The influence of wind speed and temperature on the activity of mosquitoes, expressed as harassment of caribous near Prudhoe, Alaska. (After Brown, 1975)

inland when the weather becomes cooler. The picture is complicated by the intrusion into the region of vagrant groups of caribou (Fig. 55).

In Prudhoe, just as on Spitsbergen, the reindeer is the major herbivore. Apparently the animals do not select their food according to high energy content and easy digestibility, because plants that fall into this category grow in swampy regions with mosquitoes, where reindeer prefer not to graze. But even on Spitsbergen, where there are relatively few mosquitoes, the reindeer seldom choose such food, so that apparently they rarely enter swampy regions even in other places. As a consequence, during the greater part of the summer the lactating females have a negative energy balance due to the inadequate energy content of the food.

3. High-Arctic Continental: the Canadian Archipelago

The Canadian Archipelago consists of many islands which nevertheless exhibit a typical continental climate. The surrounding seas are perpetually frozen and only in a few places is the ice broken in summer by water movements. Thus the ocean cannot exert its usual moderating influence on the temperature and the result is a continental climate with relatively strong irradiation in summer and comparatively high temperatures. Very large numbers of investigations have been carried out in this region (see Nettleship and Smith, 1975). On Ellesmere Island the vicinity of Lake Hazen was subjected to a particularly careful study in the course of an extensive investigation; and at a later date the Truelove Lowland of Devon Island was investigated within the framework of the IBP (Bliss, 1977). We have made use of data from both of these programmes in drawing up our description. Although the two areas are on a similar latitude to Spitsbergen they show very distinct differences. Plant growth on Devon Island or Ellesmere Island is comparatively small even though the number of species is high. Only in regions receiving adequate moisture during the brief growth period is there a closed covering of vegetation: large areas are almost barren. In contrast to the poverty of the flora, the fauna is rich and highly diverse. Not only are there large numbers of butterfly species, large Diptera (Tipulidae) and two regularly occurring species of bumble-bee (Oliver, 1963) which are not found on Spitsbergen, but the avifauna, too, is different (Nettleship and Maher, 1973). Whereas the Alcidae play a dominant role on Spitsbergen they are almost completely lacking on the Canadian Archipelago, where other true marine bird forms are comparatively rare. The glaucous gull, arctic tern, turnstone and king eider are the only regular visitors to the freshwater Lake Hazen. The typical avian predators of small mammals, the snowy owl and the arctic skua, are regular breeders. Wild geese can usually be considered rare visitors. Mammals are chiefly represented by lemmings, reindeer (on Lake Hazen only) and muskoxen. In the area studied on Devon Island no reindeer were encountered, Although the herbivorous snow goose is apparently found here as well as arctic hares. The most common mammalian predators are ermine and arctic fox; wolves are very rare visitors.

In the studies carried out on Lake Hazen emphasis was placed on the effect of abiotic factors on the fauna, whereas in the slightly later investigations on Devon Island an ecosystem analysis was attempted (summarized in Bliss, 1975 and 1977). The following synopsis is based upon the results of the latter:

The area of 43 km² investigated at 75° 33'N on the north-east coast of Devon Island is made up of 42% meadow, mainly with sedges; 20% dwarf shrub heath; 5% almost bare rock and 22% small bodies of water. Over 200 insect species were recorded, including 6 species of lice, 13 butterflies, 1 caddis fly, 3 beetles and 36 Hymenoptera. Nine spiders were noted. The most numerous poikilotherms are apparently the chironomids, with more than 45 species, together with the higher flies, of which more than 23 species were recorded. All of these forms take a very long time to develop. The collembole Hypogastura tullbergi, for example, requires 1–2 years to attain sexual maturity, after which it may live for another 3–4 years. Moths such as Gynaephora groenlandica and G. rossii may take up to 10 years to complete one generation, which means that population density data alone give no indication of productivity (Tables 19, 20).

Of special interest are the herbivorous butterfly and moth larvae, even if their role in the energy cycle is a very minor one. The larvae of the two Gynaeophora species mentioned above feed on the leaves of Dryas, Salix, and Saxifraga, and increase their starting weight by about 2.2 times each year. Utilization of the energy consumed is very poor (24%) since the larvae only feed for very short periods of time, but still require energy between their feeding sessions. In an average generation, 68% of the energy assimilated is lost in respiration and only 32% is assimilated in the body tissues. Another interesting point is that cold stenothermic species of this kind have a Q_{10} of 6–9, which would imply metabolic collapse at slightly higher temperatures.

The role of homeothermic animals is a larger one. Arctic hares are relatively infrequent, and only 5–10 individuals were recorded on the 43 km², probably as a result of excessive hunting in earlier times. In most years lemmings are not particularly numerous either. Their numbers usually drop by 75% in summer and only rise again in the follow winter. There is no evidence of a regularly recurring three- or four-year lemming cycle, although high changes in population density do occur. The overall low density is thought to be connected with the harsh environmental conditions. Mortality is just as high in winters with very little snow as in those with very early snowfall or an unexpectedly early thaw caused by föhn winds. The animals exhibit the usual change of biotope between meadow areas in summer, when the ground is frozen solid, and dry areas in winter.

The muskox is quite an important species here. During the period covered by the investigation its numbers rose from 254 to 271 individuals, although there were some errors in the first count and the actual increase may have been smaller. The snow goose appears to be one of the most frequently encountered birds, with 22–26 individuals, of which several pairs breed. The species is also regularly observed on a neighbouring low promontory, although the majority do not breed.

The energy flow of the area investigated is summarized in Figs. 127a, b (see also Table 21). It can be seen that homeothermic herbivores make a substantial contribution to the energy flow (approximately comparable to that found in treeless regions of the tropics and temperate latitudes; see Remmert, 1974). Although the

Table 19. The densities of invertebrate organisms in various arctic regions (individuals per m^2). About 60,000 Collembola per m^2 are reported from the Palmer peninsula. Although this is higher than all values from the Arctic, speculation as to the reason seems to be premature, especially since equally high figures have often been reported from temperate latitudes

	Wet meadow				Dry meadow			
	High Arctic	Spitsbergen	Subarctic Hardangervidda		High Arctic	Spitsbergen	Subarctic Hardangervidda	
	Devon Island		Soil	Vegetation + litter	Devon Island		Soil	Vegetation + litter
Enchytraeidae	18,600		12,000–69,000 ⌀38,000		5,800		5,000–37,000 ⌀18,000	
Rotatoria	320				110			
Tardigrada	1,160				2,800			
Copepoda	19,300				—			
Ostracoda	500				—			
Cladocera	5				—			
Acarina	3,230	22,000	25,000	6,700	4,900	21,300	95,000	5,700
Arancae	1.8	—			0.3	74		46
Collembola	8,737	243,500	33,000	6,100	18,200	20,800	62,000	6,100
Lepidoptera	—				0.2			12
Diptera Tipulidae	5				—			
Other Nematocera	7,000	2,500		24	263			42
Cyclorrapha	35				8			
Hymenoptera	1.8	20		3	8	3		16
Coleoptera	—			18	8	1		10
Hemiptera	—			28	—			450
Thysanoptera	—			44				400
Mecoptera	—			X				1

Table 20. The densities of above-ground insects on Spitsbergen and Devon Island. (Number per m^2)

	Dry meadow Devon Island	Wet meadow Devon Island	Wet meadow Spitsbergen	Dryadion Spitsbergen
Chironomidae	78	619	411	13
Sciaridae	3	6	—	22
Mycetophilidae	0.3	0.4	8	2
Empididae	1	—	—	—
Dolichopodidae	0	0.2	—	—
Muscoidea	4	11	9	1
Tenthrediniae	0.4	0.4	—	—
Ichneumonidae	1	3	12	11
Araneae	0.5	1.5	Not det.	25

Table 21. A summary of the energy flow through invertebrates in the region investigated by Bliss and co-workers on Devon Island. (Ryan in Bliss, 1978)[a]

Invertebrate Taxon	Raised beach			Meadow		
	Maximum lifespan (seasons)	Respiration	Production	Maximum lifespan (seasons)	Respiration	Production
Protozoa		?	?	0.17	17,500	8,500
Rotifera	0.3	0.1	0.1	0.5	2.6	3.2
Nematoda		4,589	3,336		561	408
Enchytraeidae	1.4	1,616	1,930	2.8	2,563	3,104
Tardigrada	0.7	151	193	1.3	13.9	18.4
Crustacea		0	0	2.0	1,076	410
Acarina	34	106	61	68	57	35
Araneida	8	12	23	4.1	0.8	1.2
Collembola	3.5	552	736	7.0	143	107
Coleoptera		0	0		0	0
Lepidoptera	3.1	16	13		0	0
Diptera	2.0	234	220	4.3	1,362	1,256
Hymenoptera						
Symphyta	4.0	21	15	7	25	19
Apocrita	2.3	6.0	2.6	5	9.6	4.2
Total		7,303	6,530		23,313	13,866

[a] Energy values are cal m^{-2}

contribution of ectothermic herbivores is very much lower it is relatively high as compared with the situation on Spitsbergen.

Unfortunately we do not know enough about either the ecology or the ecological significance of the vast numbers of geese in the Canadian arctic. Their colonies appear to be restricted to islands in lakes and estuaries as a result of predation by mammals. Avian predators seem to play an insignificant role in their population biology (skuas, glaucous gulls, and snowy owls). There appear to be no detailed studies on the primary production or of the biomass of these islands, nor on

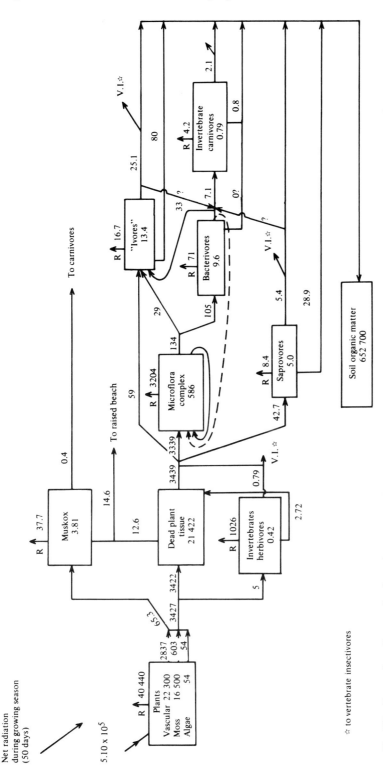

Fig. 127a. Energy flow diagram for the sedge-moss meadow system. Standing crop (*boxes*) and energy flow (*arrows*) are expressed in kJ m^{-2}. Respiration (R) is given for all components along with energy estimates for rejecta. Ivores are fungal, bacteria, and protozoa feeders, insectivores are seed and insect feeding birds. (Bliss, 1977)

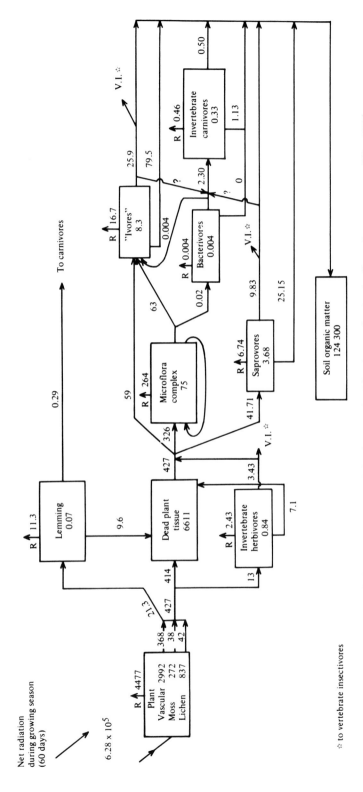

Fig. 127b. Energy flow diagram for the beach ridge (crest and slope) system. See Fig. 127a caption for details on the components. (Bliss, 1977)

☆ to vertebrate insectivores

Fig. 128. Lesser snow goose colony at McConnell River, NWT Canada, showing the occupied nests, June 18th and 19th, 1973; and the analysed photographic coverage. The numbers indicate average densities in nests/km² for the nesting areas within photo frames. (Kerbes, 1975). The whole nesting area of this colony is 90 km² with 5900 nests

their carrying capacity or on possible competition between the species or on competition with mammalian herbivores (caribou). The general appearance of these islands with their rather densely nesting geese (>1500/km²; nesting territory down to 7 m²) very much resembles that of the colonies in central Iceland (Figs. 128, 139; cf. p. 203ff.; Ryder, 1967; Kerbes, 1975), where an alternative population cycle with whooper swans is possible.

4. High-Arctic Oceanic: Svalbard or Spitsbergen

No other place on earth offers such a safe approach to within 1000 km of the North Pole, and nowhere else at these latitudes is it possible to pass both winter and summer under such tolerable conditions. Not surprisingly, it was here that modern Man first pushed so far north (William Barents), and here, at the beginning of modern times, that the first large-scale slaughter of arctic animals took place. In the seventeenth century there was a flourishing Dutch whaling town, Smeerenburg, on the west coast of Spitsbergen, with a summer population at times numbering up to 30,000 inhabitants. Due to the accessibility of the region the source of wealth in the form of whales and large seals was practically exhausted within a few decades. Later, Russian hermits came to live on Spitsbergen, supporting themselves mainly by fur-trapping. The end of the nineteenth century saw the arrival of Norwegian fur-trappers and the beginning of scientific research, which achieved its first zoological climax in Alexander Koenig's "Avifauna Spitsbergensis". In the 1920s Sumerhayes and Elton, within the framework of the Oxford expedition, attempted

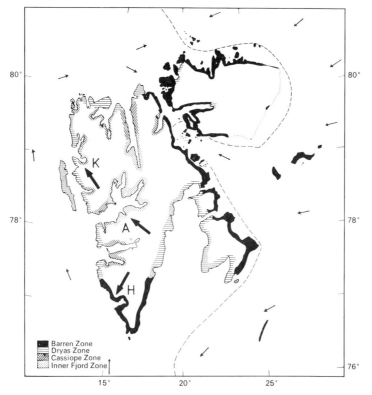

Fig. 129a. Ecological regions of Svalbard (Spitsbergen). According to Summerhayes and Elton (1928) from Remmert (1966). The area with the richest flora and fauna is the inner fjord zone, including the town of Longyearben (cf. also Fig. 118)

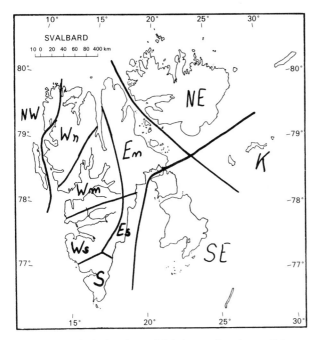

Fig. 129b. Ecological regions of Spitsbergen based on soil factors, as manifest in the occurrence of different plants. (Rønning, 1961)

the first ecological description, which is still of remarkable validity to-day (Fig. 129). The most favourable situation for both fauna and flora is offered by the central part of the main island (inner fjord zone), where the summers are relatively warm and the winters relatively cold. It is succeeded by the Cassiope zone, which in turn gives way to the Dryas zone. The latter zone is still well developed even in the icy Hornsund region at the end of the Wahlenberg fjord. The coldest ice-free parts of Spitsbergen are occupied by the barren zone, in present-day usage termed "arctic desert".

The group of islands is reached by the cold polar stream from the east, and the warm Gulf Stream coming from the south-west. Therefore the barren zone is mainly found on the eastern part of the island whilst in the west there are relatively warm areas, particularly in the north-west, since after rounding the south cape the polar stream flows northwards along the west coast for a short distance. The eastern portions of the island group may even be ice-covered in summer.

These four zones are fairly well characterized on the basis of their plant and animal species. However, this climatic zonation is modulated by geological factors, which has led to another description of Spitsbergen by Rønning (1962; see Fig. 129). Relatively sensitive plants can thrive in a climatically unfavourable habitat if a good supply of nutrients is available, i.e., on particularly good underlying rock. The two systems are not contradictory: Rønning (1963), too, describes a particularly rich flora in the central parts of the island group, corresponding on the whole with the inner fjord zone of Sumerhayes and Elton.

Fig. 130a, b

Fig. 130a-d. Tundra types on Spitsbergen. **a** Inner fjord zone; **b** Dryas zone; **c** barren zone; **d** high plateau in the inner fjord zone

Due to the meeting of cold and warm oceanic currents incoming radiation is relatively low in many places. It is highest in very cold areas and lowest on the west coast and inland regions. Thus the highest summer temperatures are still comparatively low and reach 10° C on only a few days each year. Southerly slopes in wind shadow are the only places where, on very rare occasions, the temperature climbs to 20° C.

A further consequence of the confluence of warm and cold oceanic currents—as in the case of Bear Islands—is a very rich marine birdlife. These birds and the dung provided by their droppings (some of the colonies are far inland) constitute an important component of the tundra ecosystem of Svalbard and one that has so far not been satisfactorily analysed.

The climatic zones on Spitsbergen are also modulated by differences in altitude. The many high plateaus of the inner fjord zone have a flora and fauna largely corresponding to that of the wet moss tundra that covers large areas of the transitional region between Dryas zone and barren zone at sea level, for example on the north coast of Spitsbergen.

The archipelago can thus be considered as one large ecosystem. In the following, an attempt will be made to give an overall picture of the system, drawing on studies that have largely been carried out in the inner fjord zone.

Due to the low temperatures and the small amount of sunshine, poikilothermic herbivores play almost no part in the ecosystem. In the inner fjord zone, in the Cassiope zone and the immediately adjoining parts of the Dryas zone only 3 species of aphids, 2–3 species of sawfly, 3 beetles and (perhaps) a small butterfly (erratic and rare occurrences of Plutella maculipennis) can be found. The only herbivorous insects that are encountered with anything like reasonable certainty are the sawflies (whose larvae are often found in three size groups, and thus take at least three years to develop) and aphids in the inner fjord zone. Here too a beetle species, Rhynchaenus flagellum, is encountered, although traces of its feeding are seldom detectable. Thus the invertebrate fauna is almost confined to relatively small decomposers of the litter and to predators (parasites).

Repeated attempts have been made to explain the peculiar composition of the fauna of Spitsbergen on the basis of insularity and historical factors. A glance at Table 9 is instructive: in the region of Lake Hazen, which is equally far north, there are 15 Lepidoptera species, whereas only one occurs on Spitsbergen (regularly?). This could in fact be due to its island character, but, on the other hand, Spitsbergen has five times as many Collembola as Lake Hazen. Since Collembola have far greater dispersal problems the ecological explanation is the more convincing. A further difference between the faunistic composition of arctic regions and those farther south is the degree of parasitization of the animals. Although the elevated moors and grasslands of the British Isles are reminiscent of arctic tundra they have scarcely any parasitic forms (for example parasitic ichneumonids), although these are invariably common in all genuine tundra regions and represent a major mechanism in controlling the insect populations. A survey is given in Tables 8, 18, 19.

The density and species spectrum of any biotope are highly dependent upon the degree of exposure of the site investigated (Figs. 84, 131), and upon the water supplies. The significance of all these animals within the energy cycle is mainly to be seen in their role as decomposers of the litter. More exact studies on this aspect have recently been carried out within the framework of the MaB project of the University of Trondheim. As regards number of individuals and biomass, the herbivores make no significant quantitative contribution to the invertebrate fauna on Spitsbergen. They account for less than one-half of one percent of the total catch in any trap. Obviously the role of poikilothermic herbivores here is a much smaller one than in temperate and tropical latitudes, where they account for a considerably higher proportion of all catches.

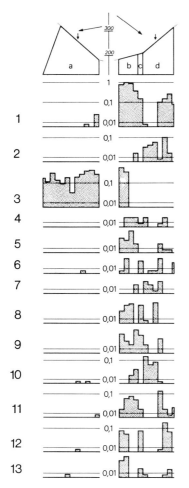

Fig. 131. Comparison between a southerly and a northerly slope in the inner fjord zone of Spitsbergen. *Above* schematic section through the slopes of the Endalen, to scale. *Thin arrows* incidence of the sun's rays on 21 June. *Thick arrows* position of temperature-recording instruments. Altitude is given in m above NN. *a* and *d* steep slopes; *b* Cassiope-Dryas heath; *c* Equisetum belt. *1* Hilaira glacialis, 449 animals; *2* Meioneta nigripes, 33 animals; *3* Collinsia spetsbergensis, 413 animals; *4* Cornicularia clavicornis, 12 animals; *5* Conigerella borealis, 22 animals; *6* Acroptena frontata, 16 animals; *7* Epistrophe tarsata, 11 animals; *8* Rhamphomya caudata, 31 animals; *9* Rhynchaenus flagellum, 35 animals; *10* Aphidoidea, 45 animals; *11* Coelosia tenella, 49 animals; *12* Mycomyia species, 23 animals; *13* Atlodia species, 13 animals. Numbers of animals have been converted to captures per day and trap, and the lines represent 0.01, 0.1 and 1 animal per trap and day at the different sites (Spitsbergen; Hinz, 1976)

Insects undoubtedly play an appreciable role in flower pollination on Spitsbergen (Fig. 132). A comprehensive analysis by Hoeg (1932) showed that especially in many centrally situated localities on Spitsbergen the nature of the vegetation suggest the presence of a rich insect fauna, and in particular of large numbers of bees and butterflies. But in fact, both groups are totally missing (with the exception of the rare and unpredictable occurence of single specimens of the moth Plutella maculipennis). Almost every species that is designed for insect pollination (79 according to Hoeg) can also reproduce in other ways, and Hoeg lists the following alternatives:

1. Completely self-pollinating: 55 species.
2. Self-pollination possible, but ± difficult. ± often pollinated by insects (Stellaria longipes—reproduces almost exclusively by vegetative means; Cerastium regelii; 3 or 5 Ranunculus species; Rhodiola?; Saxifraga hieraciifolia, S. aizoides, S. hirculus; Dryas): 12 species at most.

Fig. 132. An empidid (Rhamphomyia caudatum) in a Dryas blossom (Spitsbergen)

3. Probably always pollinated by insects (Silene acaulis): 1 species.
4. Probably tend to anemophily (3 Salix species, Empetrum): 4 species.
5. Flowers few or none; regularly reproduce by vegetative means (Polygonum viviparum, Cardamine pratensis, Saxifraga comosa, S. cernua; probably also Stellaria longipes, Saxifraga flagellaris and others): 4 species.
6. Parthenogenetic (Taraxacum): 2 species.
7. Pollination unknown; at the present time fruits never (?) ripen (Rubus): 1 species.

In the central part of Spitsbergen, hoverflies, empidids, various other species of flies and Nematocera (chironomids, mycetophilids, sciarids) are frequently seen inside blossoms. Although the empidids and hoverflies are absent from the cooler coastal regions there is no reason why pollination should not be possible. However, even the 55 species that are well equipped for self-pollination are usually pollinated by other means. From the long-term point of view this is the only way in which adaptation to changing environmental conditions by continual selection of the most suitable individuals is possible. It seems justifiable to assume that all species that are equipped with a large display apparatus are at least occasionally pollinated by outside agents, and that this ensures their continued existence on Spitsbergen. Probably about half of the vascular plants of Spitsbergen (the other half consists of grasses and sedges, broadly speaking, and is thus wind-pollinated) are therefore sooner or later dependent upon insects for pollination.

The regularly-occurring short-term cycles that are so familiar from other arctic regions do not occur on Spitsbergen. Small rodents are totally lacking and the ptarmigans apparently oscillate at irregular intervals and not in the usual four- or

Table 22. Breeding birds of Spitsbergen and at Canadian stations

	Spitsbergen	Truelove Lowland (Devon)	Lake Hazen (Ellesmere)
Gavia stellata	×	×	×
Fulmarus glacialis	×	—	—
Clangula hyemalis	×	×	×
Somateria mollissima	×	×	×
Somateria spectabilis	×	×	×
Anser brachyrhynchos	×	—	—
Chen caerulescens	—	×	×
Branta bernicla	×	—	—
Branta leucopsis	×	—	—
Lagopus mutus	×	×	×
Charadrius hiaticula	×	×	×
Arenaria interpres	×	×	×
Calidris maritima	×	—	—
Calidris canutus	×	—	×
Crocethia alba	×	×	×
Erolia bairdi	—	×	×
E. fusicollis	—	×	—
Squatarola squatarola	—	×	—
Phalaropus fulicarius	×	×	×
Phalaropus lobatus	(×)	—	—
Stercorarius parasiticus	×	×	—
Stercorarius longicaudus	(×)	×	×
Larus (Pagophila) eburnea	×	—	—
Larus marinus	×	—	—
Larus argentatus	×	—	—
Larus hyperboreus	×	×	×
Larus (Xema) sabinii	×	—	—
Rissa tridactyla	×	—	—
Sterna paradisea	×	×	×
Plautus alle	×	—	—
Uria lomvia	×	—	—
Cepphus grylle	×	—	—
Fratercula arctica	×	—	—
Falco peregrinus	—	×	—
Falco rusticolus	—	—	×
Nyctaea scandiaca	—	(×)	×
Calcarius lapponicus	—	×	×
Oenanthe oenanthe	(×)	—	—
Plectrophenax nivalis	×	×	×
Acanthis hornemanni	—	—	×

nine-year cycles (Fig. 63b). Catches of insects made in formalin traps in two successive years (1973 and 1974) in the same locality were smaller in 1974, but this must have been directly connected with the prevailing weather. Irregular oscillations seem to be recognizable in the breeding of marine birds (Lövenskiold, 1964), although nothing is known about the exact periodicity or the underlying causes (Table 22).

Fig. 133a-d. The marine birds chiefly responsible for the eutrophication of the tundra on Spitsbergen: **a** Plautus alle; **b** Uria lomvia; **c** Fulmarus glacialis; **d** Fulmarus at sea

Fig. 133d

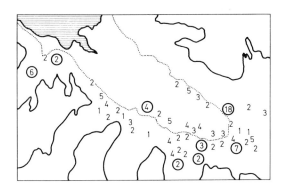

Fig. 134. Adventdal in the inner fjord zone of Spitsbergen and the distribution of the reindeer observed here on 2 July 1974. *Encircled figures* numbers of solitary reindeer. (After Pöhlmann, 1975)

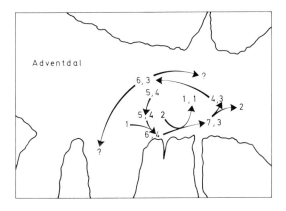

Fig. 135. Behaviour of a group of reindeer observed throughout one week in Adventdal. n, x number of adults and calves respectively. (Pöhlmann, 1975)

On the other hand, it seems possible that the reindeer populations on the entire archipelago oscillate, and that this is responsible for corresponding cycles in the vegetation. At present, there are about 7000 to 10,000 reindeer on Spitsbergen (of a very characteristic sub-species), 20,000 geese (pink-footed goose, brent goose, barnacle goose) and between 20,000 and 200,000 ptarmigans (Fig. 133). The muskoxen seem to have died out following a number of oceanic years. Compared with the other species the reindeer occupies a special position and has been most thoroughly investigated. From analysis of reindeer excreta collected over large areas it has been possible to estimate their total consumption of plant substance. Studies of this nature are only possible if several conditions are satisfactorily fulfilled:

1. The daily production of excrement in relation to food intake has to be determined in laboratory experiments. Although a relatively large number of such studies has been carried out the results show a certain amount of disparity.
2. Remineralization of the excrement must take place slowly, which is in fact the case. From the number of reindeer and the weight of excrement collected per unit area, it can be concluded that about 30 years are required for remineralization of reindeer excrement on Spitsbergen.
3. It is essential for the reindeer to be relatively evenly distributed over the tundra. The characteristically large herds of caribou in North America and of reindeer in Scandinavia render studies of this kind impossible. On Spitsbergen, however, the reindeer form small groups of 2–3 animals, consisting either of adult bucks, sub-adult animals or females with their calves (often with a yearling, probably the calf of the foregoing year); these groups are fairly evenly distributed over the tundra. Only the very wet regions near rivers and the high fjell plateau are sparsely populated. The composition of the groups alters frequently (Figs. 134, 135; Pöhlmann, 1967a, b).

The results show that the collection of excrement in test areas from which all excreta had been cleared in the previous year can provide an accurate indication of the annual production of faeces. Since the number of reindeer is known, it is possible to calculate the faeces production per day and animal, and to relate this to primary production (Brzoska, 1976). The highest annual aboveground production is yielded by meadows with Dupontia fisheri (235 g per m²), Eriophorum scheuchzeri (408 g per m²), Alopecurus alpinus (227 g per m²) and Poa arctica vivipera (421 g per m²). Since these areas are very wet they are not intensively grazed by the reindeer, at least not in summer. The animals concentrate to a much greater extent on the drier areas with Cassiope, Salix and Dryas, where productivity is much smaller. Although the aboveground biomass of such areas is almost 300 g dry substance, only 10% of this can be regarded as annual production. Large patches are totally devoid of vegetation, and the different plant communities each cover different proportions of the whole. Bearing all of these factors in mind it can be concluded that at the present density of about 10 animals per km² the reindeer consume about 10%–15% of the annual aboveground biomass, which is a very high value. The question is whether this represents a normal state of affairs.

Apparently in about 1900 there were large numbers of reindeer on Spitsbergen, and then the population collapsed, probably due to excessive hunting, and the

High-Arctic Oceanic: Svalbard or Spitsbergen

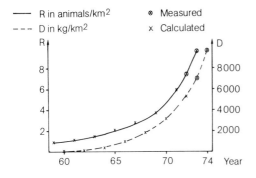

Fig. 136. Increase in reindeer population in Adventdal, Spitsbergen (*R*) and accumulation of reindeer excrements (*E*) in the same area. The total number of reindeer on Spitsbergen rose correspondingly. (Pöhlmann, 1975)

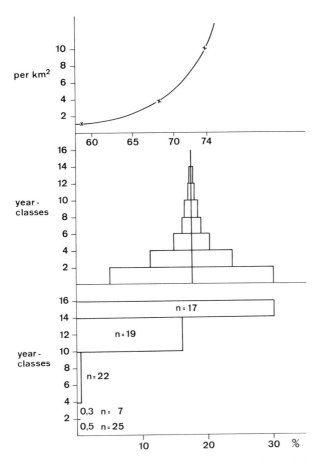

Fig. 137. The development of the reindeer population in Adventdal, Spitsbergen from 1958–1974 (*top*). *Centre*: the age pyramid calculated from the curve; *bottom*: the mortality of the various age classes (dead individuals found, n, as % of the 1974 population multiplied by 100) calculated from the age pyramid. The mortality is relatively high in the first year of life, drops sharply after this and only at an advanced age rises drastically once more. (After data of Gossow, 1972 and Pöhlmann, 1975)

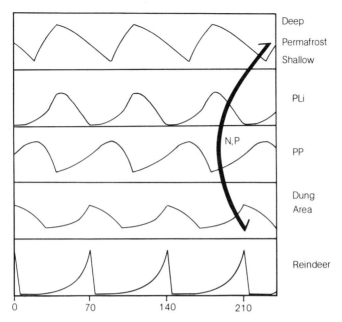

Fig. 138. Model illustrating the effect of a single factor. If a reindeer population exhibits large oscillations, as seems likely on Spitsbergen, this has an effect on the thawing of the permafrost soil in summer (*top*), on the productivity of the lichens (*PLi*), on the aboveground production of the green plants (*PP*), on the quantity of non-decomposed excrement per unit area and on the composition of the plant communities. Excrement per unit area and depth of the thawed frost soil jointly influence the availability of important minerals (*N*, *P*) to the plants. Although confirmation still has to be obtained some investigations show that the model very probably represents the true state of affairs. (Remmert, 1980)

Spitsbergen reindeer became almost extinct. Recovery was extremely slow, but eventually the numbers rose to their present level (Figs. 136, 137). The answer to the question as to whether prior to 1900 numbers were as high as today is definitely in the negative. The clue was provided by the finding of a subfossil layer of excrement in the ground, clearly distinguishable from those above and below it. This would suggest a short-term increase in numbers. On the basis of carbon particles the layer could be dated to about 1900, and can be regarded as indicating a short-lasting population explosion at that time. Prior to this the reindeer was apparently also a rare animal, even without Man's helping hand. In all probability reindeer populations are subject to very pronounced cycles with a period length of from 50 to 100 years. The animals can even multiply to a density of 15–20 individuals per km², at which point they completely exhaust the lichens that constitute their winter food. The lichens, being very slow-growing organisms, almost totally disappear from the system, the reindeer population collapses as a result, the lichens can then recover and the reindeer population recovers once more. A whole series of related events takes place more or less in parallel with such cycles. A very strong state of competition exists between vascular plants and lichens: if the lichens are over-exploited the vascular plants are able to spread so that the production of suitable summer food is at its highest just at the time when the reindeer population collapses.

During the reindeer minimum, when lichen production is again high, the sun's rays are able to penetrate further into the ground than at times when there is a covering of vascular plants. This means that the permafrost ground thaws out to a deeper level and releases larger quantities of nutrients which are beneficial not only to the lichens but to the plants that succeed them, provided that the competition presented by the lichens is removed by the reindeer (Fig. 138).

The hypothesis of reindeer-lichen cycles is based upon studies made on Spitsbergen, and on observations from south Norway, where the reindeer population, at least nowadays in the absence of natural enemies, tends to increase and to overgraze the lichens, which are then replaced by vascular plants. When, as a result of the lack of winter food, reindeer numbers have sunk to a low level, the lichens begin to spread again.

Recent investigations have shown that the Peary reindeer at no time of year consumes large quantities of lichens, which illustrates the high degree of variability between different reindeer populations, and may lead to a substantial revision of our predictions for Spitsbergen (Shank et al., 1978).

A sudden increase in numbers of a species must greatly affect the age pyramid. The younger members of the population dominate the picture to an unusual degree. The pyramid constructed from the curve of today's population figures is shown in Fig. 137. The mortality within any one age group can be found by comparing the findings of dead animals of this group with the calculated size of the group as indicated in the age pyramid. A good account of the finding of dead animals is given by Gossow and Thorbjørnsen (1972).

The frequent comparisons of Spitsbergen with central Iceland usually overlook the fact that whereas Iceland is situated immediately to the south of the Arctic Circle, Spitsbergen is little more than 1000 km from the North Pole. The much stronger insolation and the very much larger daily fluctuations in temperature in central Iceland are clearly reflected in its flora and fauna. In the course of the ecosystem project in Thjorsarver in central Iceland no less than 12 species of butterfly and moth, 3 sawflies and one bumble-bee were found, as well as one land snail. Some Lepidoptera, like Xanthorhoe munitata for example, are extraordinarily common. There are no culicids but only simuliids on Iceland. Thus the situation is a very different one from that seen on Spitsbergen. A special feature of the central Icelandic tundra is the complete absence of mammals, and the only herbivores of any importance are the wild geese (Anser brachyrhynochos). It is possible that there is a long-term cycle involving sedges, Characeae and wild geese on the one hand, and the whooper swan and cotton grass on the other, as the alternately dominating organisms in this oasis in the volcanic desert of central Iceland (see p. 189; Fig. 139; Gardasson, 1975, 1976, 1979).

5. Arctic Lakes

In arctic lakes, between 1 and 170 mg carbon per square meter and day are incorporated into organic biomass, which amounts to 1–35 g C per m² annually

Fig. 139. a A view of the Thjorsarver colony of the pink-footed goose in central Iceland. (Gardasson, 1975). b A view into a colony of Ross' geese in nothern Canada. (J. Ryder phot.). c Aerial view of typical habitat of Ross' geese with a breeding colony on an island. Northern Canada, 3.7.; 66°N. (J. Ryder phot.)

(Likens, 1975; Table 23). In actual fact the possible range is much larger. The fauna is extremely varied. Unfortunately, many of the very important lakes of the Arctic have only been unsatisfactorily investigated (e.g., Torneträsk, Inari See, Lake Hazen), so that comparisons are scarcely possible. From the zoologist's point of view, however, this would be a very attractive undertaking: after the ice of the glacial epoch had melted many arctic regions rose, with the result that in numerous places freshwater populations of marine animals developed in complete independence of one another. Examples are provided by the seals (ringed seal in Lake Saimaa in Finland, harbour seals in Labrador), and by a wealth of primary marine animals (Mesidothea, Pontoporeia, Mysis). Zoologists ought to know more about the role of such "foreign" organisms in the limnic fauna. The presence of such marine elements in many freshwater lakes throughout the Arctic is responsible for the astonishing diversity of their fauna and for extremely large differences between the fauna of the various lakes, even those that are relatively close together. Probably the only characteristic common to all arctic lakes is the absence of the daily vertical migration of the plankton that is elsewhere so characteristic, both in marine and freshwaters.

The largest lake of the Canadian Archipelago, Lake Nettling, on Baffin Island, is situated on the Arctic Circle. It has a surface area of 5,699 km^2, and a mean depth of 25–55 m. The lake is highly oligotrophic, and visibility extends to a depth of 11–15 m in summer (minimum 1.1, maximum 21.3 m).

It is frozen over each winter and in May the ice is from 1.5 to 1.9 metres thick. When the thaw begins, a layer of melt water covers the ice to a depth of about 10 cm, and may even freeze again completely in June. Not until the end of July does the ice break to such an extent that the lake can no longer be crossed on foot, and not until the beginning of August is it ice-free. The pH of the water is in the range of

Table 23. A survey of the normal net primary productivity of various freshwater ecosystems. (Lieth in Whittaker, 1975)

Water system	mg C/m^2/day	g C/m^2/year[a]
Tropical lakes	100–7,600	30–2,500
Temperate lakes	5–3,600	2–950[b]
Arctic lakes	1–170	< 1–35
Antarctic lakes	1–35	1–10
Alpine lakes	1–450	< 1–100
Temperature rivers	< 1–3,000	< 1–650
Tropical rivers	< 1–?	1–1,000?

[a] In most cases, averaged over estimated "growing season"
[b] Naturally eutrophic lakes may reach a maximum of 450

6.8–7. From the list of animals given in Table 26 it can be seen that the benthos is mainly made up of chironomid larvae. In this it differs greatly from other large lakes of the Canadian Archipelago, in which amphipods (in fact a species of marine origin, Pontoporeia affinis) are generally more numerous than chironomids. As is usual in cold regions copepods make up the main bulk of the zooplankton. Thus, on the whole, the fauna is poor, and attention is repeatedly drawn to the paucity of amphipods and molluscs. This becomes very obvious if the list is compared with that compiled by Müller (1974) for Messaure, Sweden, at almost exactly the same latitude (see Tables 13 and 14). In the Messaure vicinity 12 species of fish can be found in freshwater (Table 15) despite the fact that the freshwater lakes in this region are ice-covered almost as long as Nettilling Lake.

So far the basic ecological problems have not even been afforded preliminary attention. We know nothing about interconnections between the various species, we have no information concerning energy flow, and we are not in a position to make comparisons with the lakes of temperate latitudes. In elegant experiments Fisher and Rosin (1968) demonstrated that almost 100% of the chironomid pupae rising to the surface from the bottom of an illuminated body of water are devoured by fish, whereas those rising at twilight or in the dark are scarcely endangered at all. This explains why the majority of aquatic chironomids of temperate and tropical latitudes hatch from the pupae towards sunset or after. In High-Arctic waters, too, chironomids hatch at fixed times. However, during the arctic summer the body of water through which they have to rise is light enough for them to be seen by predatory fish. This clearly brings up a number of questions that still have to be answered. The same applies to the problems connected with the diurnal vertical migrations of planktonic organisms. McAllister showed that a higher productivity of a body of water is guaranteed if the planktonic organisms feed at night, when the planktonic algae upon which they feed are unable to synthesize, than if they feed all around the clock, i.e., including the times when the phytoplankton are assimilating CO_2. Sinking to colder water layers during the non-feeding hours means a reduction in energy consumed in metabolism. Thus, theoretically, a diurnal vertical migration of animal plankton in arctic seas and lakes could lead to higher productivity. It was pointed out at the beginning of this book that the zeitgebers are adequate for diurnal migrations of this nature, and Haney demonstrated that the plankton in

Arctic Lakes

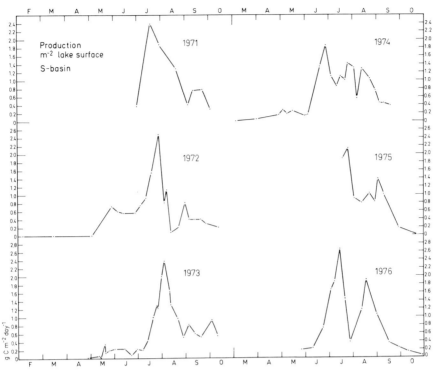

Fig. 140. Primary production of phytoplankton of Myvatn (Iceland) from 1971 to 1974. (Jonasson, 1979)

High-Arctic lakes also feed according to a strict diurnal rhythm. Therefore the absence of daily vertical migrations must have a different explanation.

The only investigations on a cold lake that might in any way throw some light upon these problems have been carried out in the Alps (Vorderer Finstertaler See; see Pechlaner, 1971). However, due to the southerly position of the lake and the completely different conditions entailed (regular diurnal alternation of light and dark, diurnal vertical migrations of plankton) it cannot be used as a basis for comparison.

Although at almost the same latitude as Nettilling Lake, Myvatn in north Iceland, whose bird-life makes it one of the most famous lakes of the world, is entirely different. Normally the lake has a thick covering of ice from the beginning of November until mid-May, despite the fact that some of its inflow comes from hot springs. The lake itself consists of a northern and a southern basin, with a total surface area of 37 km² and a maximum depth of 4.2 m. Its outflow to the sea is the Laxa river, which flows northwards for 55 km, carrying approximately 33 m³ per second. The lake receives its water from springs, chiefly on its eastern shore. The springs feeding the southern basin are cold (5°–6° C), whereas those supplying the northern basin have temperatures of up to 23° C. In summer the temperature of the lake water may rise to as much as 15°–18° C. The water is rich in plant nutrients, which accounts for the amazingly high productivity, with values exceeding by

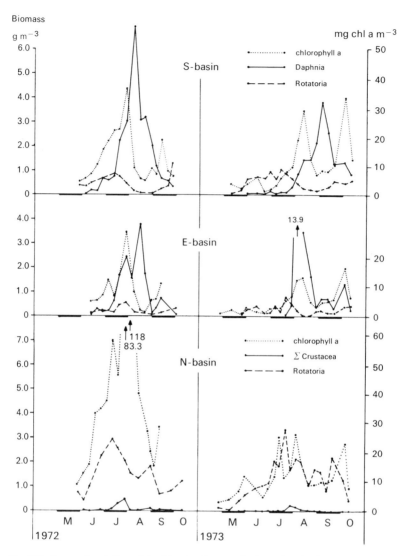

Fig. 141. Seasonal variation of plankton development in the various basins of Myvatn. (Jonasson, 1979)

orders of magnitude those normally given for arctic lakes. The primary production of the phytoplankton in the southern basin may amount to as much as $110°$g C per m² and year (about the same as eutrophic lakes of temperate latitudes). Production already begins in March, when the lake is still covered by ice. This primary production is succeeded by a mass development of planktonic animals (Figs. 140–142), with copepods and rotifers playing a major role in spring, and Daphnia appearing in vast numbers in June and July. The following succession can be observed: relatively large rotifers (Polyarthra, Synchaeta) and Nauplius larvae dominating in spring, together with diatoms and, slightly later, with

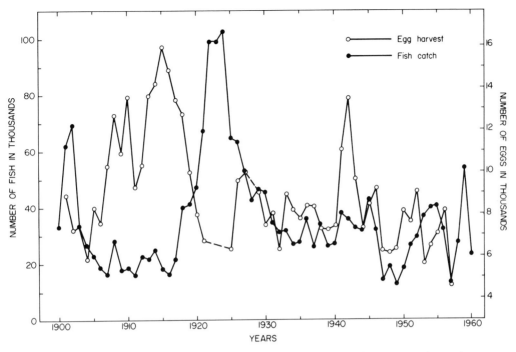

Fig. 142. Total annual fish catch (Salvelinus alpinus and Salmo trutta) in Myvatn and annual egg harvest of the farm Grimsstadir. (There are 14 farms harvesting eggs in the Myvatn area). This figure shows better than words the high productivity of the lake. (Gudmundsson in Jonasson, 1979)

chrysomonadids. When these have reached their peak and diatom multiplication has been arrested by bacteria, smaller rotifers (Keratella) become very numerous. At the end of May or beginning of June, Daphnia hatches and achieves its greatest productivity, together with the algal flowering of Anabaena. Daphnia and the smaller rotifers dominate, and the algal bloom collapses in August. The spring rotifers now become more numerous again and late autumn copepods (Cyclops) and the predatory rotifer Asplanchna appear once more. A total production of 30–100 g animal plankton (fresh weight) is achieved, whereby Daphnia is mainly responsible for the annual fluctuations in production figures. The bottom fauna consists almost entirely of chironomids, although their numbers have considerably declined in recent years (Fig. 143). In addition, copepods and Cladocera are found.

Thus the three fish species of the lake (Salvelinus alpinus, Gasterosteus aculeatus and Salmo trutta) are plentifully provided with food. The stomach contents of the char as a rule reflected very accurately the distribution of the various food species at any one time. The chief component was chironomids (usually pupae); Lepidurus was only consumed in summer after the chironomids had hatched. Sticklebacks and the snail Lymnaea are only consumed if the principal food species are scarce.

Such a high production makes possible the high density and large species diversity of aquatic birds for which the lake is world-famous. Apparently, numbers

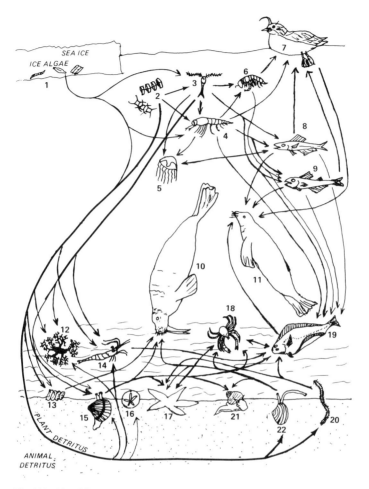

Fig. 143. Simplified representation of the Bering Sea food web. Organisms and assemblages of organisms depicted are as follows: *1* ice algae; *2* phytoplankton; *3* copepods; *4* mysids and euphausids; *5* medusae; *6* hyperiad amphipods; *7* sea birds; *8, 9* pelagic fishes; *10* walrus; *11* seals; *12* basket stars; *13* ascidians; *14* shrimps; *15* filter-feeding bivalves; *16* sand dollars; *17* sea stars; *18* crabs; *19* bottom-feeding fishes; *20* polychaetes; *21* predatory gastropods; *22* deposit-feeding bivalves. (McConnaughey and McRoy, 1979)

were greatly exaggerated in earlier reports, so that exact data on the population dynamics of the birds cannot be given. Figures published today are without exception lower than those in earlier publications so that whether the numbers are, in fact, declining cannot be stated with certainty. The species found on the lake are listed in Table 24 (Gardasson, 1975; Jonasson et al., 1977; review: Jonasson, 1979).

Thus it seems that lakes in the immediate vicinity of the Arctic Circle can vary enormously with respect to productivity, species numbers and species composition. At very high latitudes the differences are probably smaller, but again investigations giving more than a rough survey of the fauna are completely lacking.

Table 24. Breeding birds of the Myvatn region of Iceland. (Gardasson, 1979)

Species	No. of breeding pairs
Surface feeders (mostly phytobenthos, some benthic animals):	
Anas platyrhynchos	350
A. acuta	
A. strepera	
A. clypeata	
A. penelope	
A. crecca	
Diving ducks, usually feeding on zoobenthos:	
Bucephala islandica	700
Aythya marila	1,700
A. fuligula	3,000
A. ferina	
Clangula hyemalis	
Histronicus histronicus	
Melanitta nigra	
Fish-eating mergansers	
Mergus serrator	
Mergus merganser	

An amazing wealth of species is found in the northernmost lakes of Scandinavia, such as Kilpisjärvi in north-west Finland, approximately 600 km north of the Arctic Circle (69° 45'N). This is an oligotrophic body of water covering roughly 37 km². Although there are clearly fewer species at great depths than near the surface, the types of animals and number of species in the lake are far more numerous than in comparable lakes in North America, or even in Myvatn in Iceland. Whereas the poverty of Icelandic lakes can probably be put down to the insular character of the country, climatic factors are undoubtedly responsible for the differences between American and Scandinavian lakes (Tables 25, 26). The benthos of Kilpisjärvi harbours no fewer than 4 species of Cladocera, 7 Ephemeroptera, 10 Plecoptera, 2 Bryozoa, 3 Pisidium, 2 snails, 1 leech (Glossiphonia complanata), 1 amphipod (Gammarus) besides many other species (chiefly chironomids). The figures for the biomass of the bottom fauna tend to be somewhat higher than those for comparable lakes in southern Finland, which is in agreement with the findings of Thorson and Hinz. In any case, a complicated pattern of ecological relationships can be anticipated, although so far no studies have been undertaken in this direction.

It is impossible to say at the moment whether or not this elevated biomass also applies to the fish, firstly because of the large differences between the various lakes, and secondly because of the extent of human interference in every case. From a lake in northern Finland, 2.4 km in length, 0.2 km wide and 25 m deep, Lind (1974) lists 70 Rutilus rutilus and 150 Perca fluviatilis per ha, in addition to very small numbers of Lota, Esox, Coregonus, Salmo trutta, Pungitius, and Phoxinus. Per ha he reports

Table 25. The bathygraphic distribution and density of bottom-dwelling animals in Lake Kilpisjärvi, North Finland

Depth zone (m)	1–2	2–3	3–4	5–5	5–7.5	7.5–10	10–15	15–20	20–25	25–30	30–35	35–40	40–45
Vorticella sp.	–	+	+	–	+	–	+	+	–	+	–	+	–
Turbellaria	53.9	341.0	31.0	–	40.3	31.0	20.4	12.4	–	–	–	–	–
Nematoda	207.9	77.5	–	46.5	95.2	156.0	119.0	37.2	19.6	11.6	11.6	31.0	7.7
Oligochaeta	1,686.3	852.5	310.0	356.5	362.3	300.0	492.0	316.2	110.0	108.4	89.0	38.7	79.5
Cladocera	15.4	31.0	–	–	47.6	206.8	13.6	6.2	2.8	–	–	–	7.7
Copepoda	15.4	15.5	–	15.5	11.0	10.3	–	6.2	8.4	–	–	–	–
Ostracoda	–	31.0	–	–	–	5.2	–	–	–	–	–	–	–
Ephemeroptera	7.7	–	–	–	res	–	res	res	–	–	–	–	–
Trichoptera	7.7	–	–	–	–	–	3.4	≪	–	–	–	–	–
Diptera (Tipulidae)	23.1	15.5	62.0	–	7.3	10.3	3.4	–	–	–	–	–	–
Diptera (Chironomidae)	1,409.1	1,643.0	465.0	666.5	1,753.0	1,364.9	1,023.4	508.4	160.5	100.6	85.2	108.5	61.9
Hydracarina	77.0	372.0	186.0	155.0	73.2	41.4	20.4	24.8	14.1	7.7	31.0	–	61.8
Pisidiae	107.8	387.5	–	31.0	32.9	41.4	47.6	34.1	8.4	23.2	–	15.5	7.7
Gastropoda	–	46.5	31.0	15.5	7.4	5.2	–	–	–	–	–	–	–
Bryozoa	–	15.5	31.0	res	3.7	31.0	3.4	12.4	5.6	3.9	3.9	–	–
Average density	3,603.6	3,828.5	1,116.0	1,266.5	2,462.6	2,197.0	1,744.2	961.0	330.0	255.4	220.6	193.7	224.7

Table 26. The fauna of Lake Nettilling, North Canada. (Oliver, 1964)

Salvelinus alpinus	Psectrocladius colcaratus
Gastrosteus aculeatus	Parakiefferiella nigra
Pungitius pungitius	Prissocaldius sp.
	Chironomus (s. s.) sp.
Pisidium lilljeborgi	Stictochironomus rosenscholdi
P. casertanum	Micropsectra natvigi
	Sergentia coracina
Gammarus lacustris	Lauterbornia coracina
Ostracods	Oeklandia borealis
	Procadius sp.
Acalptonotus violaceus	Diaptomus minutus
Lebertia sp.	Cyclops scutifer
Neobachypoda ekmanni	Cyclops capillatus
Hygrobates foreli	Cyclops languidoides
Pseudosiamesa arctica	Daphnia longiremis
Heterotrissoclaidius subpilosus	Daphnia rosea
Cricotopus alpicola and probably	Bosmina longirostris
(imagoes found along the shore):	Chydorus sphaericus
Protanypus (?) caudatus	Holopedium gibberum
Orthocalius consobrinus	Alonella nana

a total of about 13 kg Rutilus, 9 kg perch, and 1 kg of all the remaining species together. Rutilus rutilus produces (inclusive of gonads) approximately 4.3 kg per ha and year, perch about 4 kg, and the remainder about 0.8 kg, so that, in all, 9 kg of fish are produced per ha and year. In a small pond nearby, containing only perch, the numbers were approximately 840 individuals per ha with a total weight of 31 kg. Exclusive of the gonads, production here amounted to about 6.3 kg. Taking the weight of the gonads to be the same in the pond as in the lake, a figure of 9 kg per ha and year is again obtained. Although these values are less than 50% of the average Finnish values, they considerably exceed those for High-Arctic lakes.

A completely different state of affairs is seen in lakes of the extreme High Arctic. As an example we shall consider Lake Hazen on Ellesmere Island (almost 82°N). This is a long (10 km) and deep (280 m) lake, and although at all times, even in winter, there is open water at its outlet to the sea, there are many summers in which it is not entirely free of ice. This is partly due to the low wind speeds in the region. In a good year the ice disappears at the beginning of August but the lake is frozen over once more by the end of September.

Most of the primary production apparently goes on beneath the ice. Practically no producers can be demonstrated following ice-break and productivity at this time is below the level at which it can be demonstrated by the dark-bottle method. The bottom fauna consists solely of chironomid larvae and aquatic mites. The plankton contains two rotifers (Keratella hiemalis and K. cochlearis) and a copepod (Cyclops scutifer), which support a population of lake char (Salvelinus alpinus).

None of these species manages to achieve a high population density. Only a very few individuals of K. cochlearis were found; in some years K. hiemalis is more and in some years less numerous than the copepods. As regards biomass, however, copepods always account for the major portion but since they take two years to develop productivity is correspondingly small, which is not surprising in view of the low temperatures of the lake (never above 3° C). McLaren (1964) calculated the copepods at 96 mg per m² and net production at 1 mg dry weight per m² and day. In small ponds in the same vicinity, which become very much warmer in summer, the same species are found but Cyclops scutifer completes its cycle in one year and its productivity is 15 times greater than in the large lake.

6. The Arctic Seas of the Old World

The Gulf Stream which drives water up the coast of Norway beyond the North Cape as far as Spitsbergen at 80°N makes possible the passage of shipping all the year round in practically ice-free harbours, Hammerfest and Murmansk, at a latitude of 70°N (Fig. 144). The confluence of the North Atlantic current and the cold polar current creates a region of strongly upwelling water, which leads to high production and a vast abundance of fish. Not surprisingly, scientific research in the Arctic Seas of the Old World began comparatively early. An added impetus was supplied by the prospect of the economic advantages of a connection between Europe and north Asia and Alaska via polar waters. A very different situation

Fig. 144a-e. Coasts of the Arctic. **a** Rocky coast near Tromsö (Norway); **b** rocky coast within the sphere of influence of the Gulf Stream (west Spitsbergen); **c** rocky coast on north Spitsbergen; **d** rocky coast in east Greenland, approx. on the Arctic Circle; **e** the largest bird mountain of the Arctic: Bear Island (**d, e** see p. 216)

Fig. 144b, c

Fig. 144d, e. Legend see p. 214

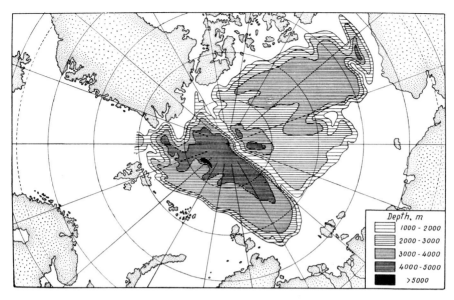

Fig. 145. Arctic basin bottom topography according to data from Soviet drifting observation stations. (Zenkewitch, 1963)

prevails in the New World: apart from occasional gaps where the ice has either collapsed or melted, there is a more or less permanent covering of ice. Thus, as we have already seen, the northern half of Greenland and the Canadian Archipelago can be considered a continent, with a purely continental climate. Serious marine research is of a very recent date here and lags far behind the research already carried out in the terrestrial sphere.

A review compiled by Zenkewitch (1963) covers marine biological investigations in the Old World Arctic, most of which were carried out in Russian marine stations (some of them long-established) and are for the larger part inaccessible to western scientists. Our description is based almost entirely on his representations, and only in a few cases has reference been made to more recent literature.

The north coast of the Soviet Union can be fairly clearly divided into a series of marine regions. The Barents Sea, bordered to the north by Spitsbergen and Franz Josefs Land, and in the east by Novaya Zemlya, is in fact still a part of the North Atlantic Ocean. The White Sea, which adjoins to the south, at times contains relatively brackish water in summer due to water from melting snow. The Kara Sea extends between Novaya Zemlya and the Seven Islands, followed by the Laptev Sea, which extends as far as the Novosibirsk Islands. The adjoining East Siberian Sea is limited by Wrangel Island, after which the Chukotske Sea stretches to the Bering Straits. To the north of these coastal seas, all of which exhibit brackish characteristics on account of their shallowness and inflow from the great Siberian rivers, is the true circumpolar Arctic Basin. It is far deeper than was previously assumed, and in some places a depth of 5000 m (Fig. 145) has been measured. The temperatures through the central portion of this region, from the Seven Islands to

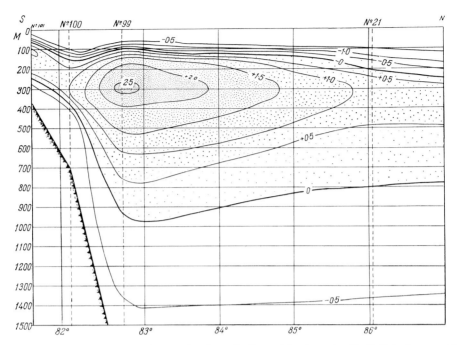

Fig. 146. Temperature curves in a section from the Seven Islands to the North Pole (i.e., roughly corresponding to a line along longitude of 100 °C. (Zenkewitch, 1963)

the North Pole, are relatively similar and consistently low (Fig. 146). The influence of the Gulf Stream declines from west to east, and consequently in summer the surface temperatures also drop markedly in the same direction. Whereas in summer the edge of the ice on Spitsbergen is only 1000 km from the pole, in Siberia it even remains on the coast, i.e., in some places to 70°N. Nowadays this presents no serious problems to ice-going ships: it is not a solid covering, but consists of ice floes and patches of pack ice, but the situation is vastly different from that at the Kola peninsula or the North Cape of Norway.

On the basis of the dominant planktonic organism it is possible to distinguish zones corresponding to the situation described above (Fig. 147). In the Barents Sea in the west the calanids of the northern seas are still predominant, whereas brackish forms and arctic oceanic forms play the major role in the east. The benthos appears to be affected by the temperature gradient to an even greater extent than the plankton. In Tromsö, for example, on the Norwegian coast about 300 km north of the Arctic Circle, practically the entire spectrum of plants and animals of the boreal tidal zone is encountered, but in the vicinity of the North Cape and to its east a slight but detectable impoverishment is noticeable. Despite this, as Fig. 148 shows, flora and fauna are still quite normal. This also holds for portions of the west coast of Spitsbergen (Magdalenenfjord, for example) which are little troubled by ice in winter, and where Fucus, Litorina rudis, Balanus balanoides, Gammarus species and Pseudalibrothus, Bryozoa and Hydrozoa are well represented. Further to the east, however, in regions regularly subjected to the action of ice, the macrofauna of

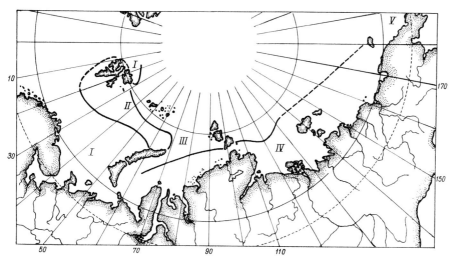

Fig. 147. Distribution of main types of zooplankton in northern seas. *I, II* Pronounced predominance of Calanus finmarchicus (90% of biomass; many boreal forms present); *III* predominance of Calanus finmarchicus (not more than 40% of total biomass) and C. hyperboreus; a considerable admixture of Atlantic forms from intermediate layer; *IV* predominance of Pseudocalanus elongatus and a selection of brackish-water forms; *V* same as *IV* but with an admixture of Pacific Ocean forms. (Zenkewitch, 1963)

the tidal region is very poor and finally disappears altogether. Only isolated barnacles (which no longer reproduce) and small Fucus thalli manage to survive in rocky clefts.

Although in the deeper zones the number of species is low, there is a distinct increase in biomass, recognizable in the larger number of individuals (Fig. 149). Only in the deep Arctic Basin itself and in the Chukotske Sea are the numbers of species and individuals extremely low. The deep zones of the Arctic Ocean do not seem to differ basically from those of other oceans, and the picture of the ocean floor as revealed by collections and photographs (e.g., Hunkins et al., 1970) differs in no way from that of other oceanic regions. As everywhere, signs of life are detectable at great depths, but in low numbers. Photographs show Holothuria, starfish, brittle stars, crinoids and traces of a variety of other animals. An interesting peculiarity of the arctic deeps are the "findlinge", which are large rocks that have been transported with the glacial ice from the mainland to the deep sea, and "dropped" by the icebergs.

The productivity upon which this fauna depends is made possible by the light of the very brief summer season. In the actual circumpolar Arctic Basin real production is practically confined to August and the first part of September, whereas in the northern part of the Barents Sea phytoplankton produces from the beginning of June until the end of October. In the southern Barents Sea phytoplankton even produces throughout the entire year, although in the winter months productivity is extremely low (Fig. 150). The phytoplankton maximum in June is succeeded by a corresponding zooplankton maximum in August, associated with marked qualitative changes in the composition of the planktonic forms (Fig.

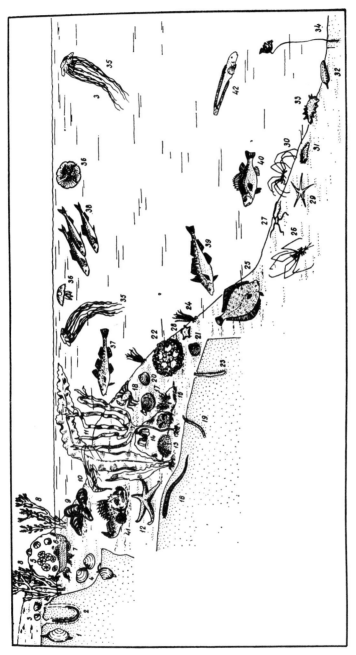

Fig. 148. Distribution of most typical representatives of Barents Sea population (Zenkevitch). *1* Mya arenaria; *2* Arenicola marina; *3* Littorina littorea; *4* Macoma baltica; *5* Balanus balanoides; *6* Littorina rudis; *7* Gammarus spp; *8* Fucus vesiculosus and Ascophyllum nodosum; *9* Mytilus edulis; *10* Pandalus borealis; *11* Laminaria saccharina and L. digitata; *12* Asterias rubens; *13* Strongylocentrotus droebachiensis; *14* Balanus crenatus; *15* Bryozoa; *16* Nephthys; *17* Pecten islandicus; *18* Hyas araneus; *19* Maldane sarsi; *20* Astarte borealis; *21* A. crenata; *22* Gorgonocephalus; *23* Phascolosoma margaritaceum; *24* Heliometra glacialis; *25* Pleureonectes platessa; *26* Munnopsis typica; *27* Ophioscolex glacialis; *28* Ctenodiscus crispatus; *29* Asterias panopla; *30* Colossendeis proboscidea; *31* Pourtaleusia jeffreisi; *32* molpadia spp; *33* Elpidia glacialis; *34* Umbellula encrinus; *35* Cyanea arctica; *36* Aurelia aurita; *37* Gadus morrhua; *38* Clupea harengus; *39* Melanogrammus aeglephinus; *40* Sebastus marinus; *41* Myoxocephalus quadricornis; *42* Lycedes spp.

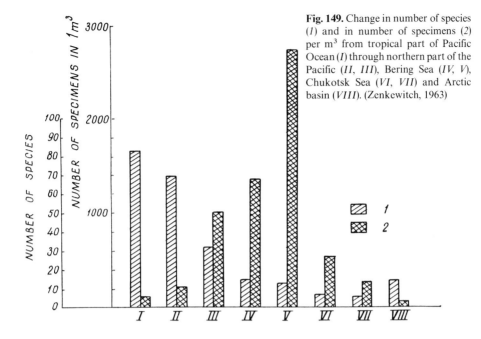

Fig. 149. Change in number of species (*1*) and in number of specimens (*2*) per m³ from tropical part of Pacific Ocean (*I*) through northern part of the Pacific (*II, III*), Bering Sea (*IV, V*), Chukotsk Sea (*VI, VII*) and Arctic basin (*VIII*). (Zenkewitch, 1963)

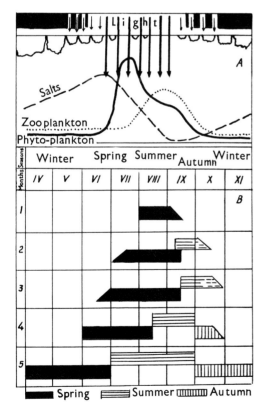

Fig. 150. Biological season of plankton. *A* General indices (Bogorov). *B* Phytoplankton development (Usachev). *1* Circumpolar part of Arctic Ocean; *2* central region of Kara Sea; *3* Laptev Sea; *4* northern part of Barents Sea; *5* south-western part of Barents Sea. (Zenkewitch, 1963)

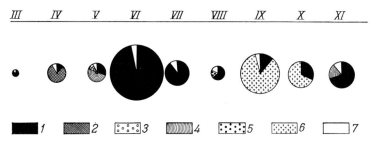

Fig. 151. Quantitative and qualitative changes (mg/m³) of zooplankton in the Motovsky Gulf in 1931 (Manteufel). *1* Calanus finnarchicus; *2* Balanus larvae; *3* Decapoda larvae; *4* Thysanoessa inermis; *5* Copepoda (summer community); *6* Limacina retroversa; *7* Varia. (Zenkewitch, 1963)

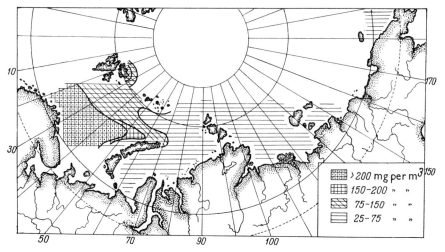

Fig. 152. Distribution of maxima of mean biomass of zooplankton of seas adjacent to eastern sector of the Arctic basin in mg/m³. (Zenkewitch, 1963)

151). The maximum resulting from the phytoplankton bloom is actually due to Calanus funmarchicus, a species that at all other times lags far behind the rest of the plankton. The gradual drop in productivity from west to east (due to temperature, ice cover and duration of growth period) is reflected in a decrease from west to east and from south to north in the maximum density of zooplankton (Fig. 152). The distribution of the maxima of the mean biomass shown in the figure is in good agreement with data for the yields of commercial fish, which are known to be considerably higher in the Barents Sea than off the Siberian coast. An important link in this nutritional chain is constituted by the herring. Most of these fish pass the winter in the deeper water layers and very few approach the surface. Calanids and Euphausicaceae behave in the same way. From February onwards the plankton and immature herrings begin to rise slowly, and to exhibit daily vertical migrations. From April onwards the planktonic animals begin to reproduce, and from May the juvenile stages sink to deeper layers. After this the adult plankton and herrings

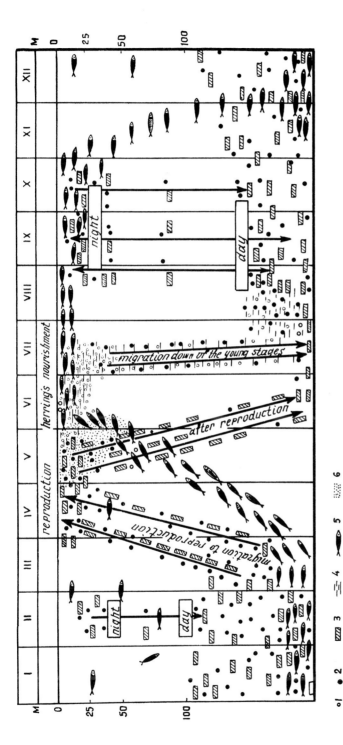

Fig. 153. Diagram of vertical migration of immature herring and of its food (Calanus and Euphausiaceae) within the region of Murman Bank (Manteufel, 1941). The direction of plankton migration is shown by the *arrows*, daily vertical plankton migrations are shown by *double-headed arrows*. *1*, *V* to *VII* Copepoda stages of Calanus finmarchicus; *2*, *I* to *V* Copepoda stages; *3* adult Thysanoessa; *4* Thysanoessa larvae; *5* herring; *6* waterbloom. (Zenkewitch, 1963)

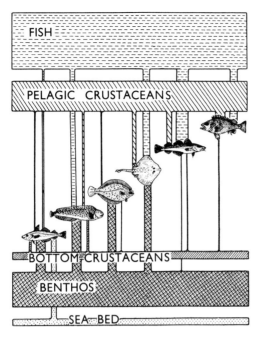

Fig. 154. Feeding habits of the chief commercial fish of the Barents Sea in order: haddock, Anarhichas, sand dab, ray, cod, sea bass. (Zenkewitch, 1963)

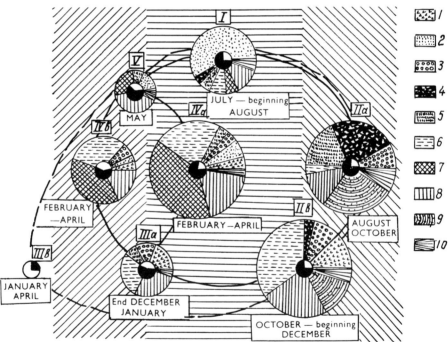

Fig. 155. Diagram of food cycle of Barents Sea cod (Zatzepin and Petrova, 1939). Contrinuous lines-immature cod; broken lines-mature cod; *I* Summer fattening; *II* main autumn feeding; *IIIa* period of lesser feeding during spawning migrations; *IIIb* during spawning; *IV* spring fattening in (*a*) central and (*b*) western fishing regions; *V* period of forced starvation. *Areas of circles* correspond to repletion indices. *1* Plankton organisms; *2* Euphausiaceae; *3* prawns; *4* bottom-living crustaceans; *5* other bottom-living animals; *6* herring; *7* caplin; *8* cod and haddock young; *9* arctic cod; *10* other fish. (Zenkewitch, 1963)

remain near the water surface of the productive zone. Daily vertical migrations now cease and do not recommence until the beginning of August. From the end of September, but for the most part at the beginning of November, plankton and fish once more sink to the deeper layers, where they remain during December and January (Fig. 153). A similar picture is revealed by the distribution of sea birds. The north coast of the Kola peninsula with its offshore islands, Bear Island and Spitsbergen, is almost completely occupied by the birds and it is difficult to tell where one colony ends and the next begins. On Novaya Zemlya, however, the colonies begin to thin out, and from this point up to the Bering Straits fulmars, for example, are apparently not encountered (Fisher, 1952).

As would be expected from the large annual fluctuations in productivity, no strict food specialization is seen in the long-lived fish that inhabit these waters all year round. They can all catch smaller fish, pelagic crustaceans and animals of the ocean floor (Fig. 154). Closer inspection reveals a distinct food cycle with alternation between various food species over the course of the year, as shown for the cod in Fig. 155. Similar food cycles can also be recognized for all other fish of commercial importance, and have also been demonstrated for Alcidae by Belopolski (Figs. 51–54).

7. Antarctica: a Comparison

The Arctic offers such a variety of form that no truly comprehensive approach has so far been attempted. The Antarctic is very much more of an entity, and reviews are available in sufficient numbers, although even here the situation is constantly changing due to the intensive research activities at present in progress (van Micghem et al., 1965; Holdgate, 1970; Dunbar, 1977; El-Sayed, 1977; Llano, 1977).

The map of the Antarctic has undergone considerable changes in recent years. Whereas earlier atlases depicted it as one large continent stretching roughly to the Polar Circle, with only the Palmer Peninsula extending further northwards in the direction of South America, we now know that the Antarctic continent in fact consists of an east and a west continental block, which have drifted together. At the point where they meet the ocean thrusts into the continent as two deep bays, the Ross Sea and the Weddell Sea, both of which are covered by a gigantic shelf, so that penetration is only possible for a very brief period in summer and then only for a very short distance. For all practical purposes, therefore, the continent retains its original form and presents a unique set of conditions. In fact, the continent is a large island situated almost exactly over the South Pole and surrounded by an ocean whose water and wind masses follow, almost unhindered, the movements dictated by the earth's rotation. Thus the continent is encircled by strong storms and water currents, and theoretically a ship, a bird or a fish might encircle it repeatedly.

These are the conditions that have prevailed for about 20 million years, and for this vast length of time the Antarctic continent and seas have been largely cut off from other continents and oceans. In addition, the very cold antarctic climate has apparently remained more or less constant throughout all these millions of years.

The picture we have drawn is grossly simplified, and obviously fresh masses of water are continually being brought in with the west wind drift and the polar current, and extensive mixing of the waters of the Antarctic and of other oceans occurs. Nevertheless, a simplification in this form has its justifications: the temperature of the surrounding seas is almost everywhere the same, and seasonal temperature fluctuations are practically unknown. The subantarctic islands lying within this windswept, stormy region are also remarkable for their very small daily and annual temperature fluctuations (Fig. 3). Temperatures are invariably low, and at the Antarctic Convergence values above 3 °C are seldom attained.

On the continent the extreme cold renders life on land almost impossible. In the Antarctic all forms of life depend to a far greater degree upon the sea than in Arctic regions, and we shall therefore deal with this aspect first of all. The following remarks are based largely upon the description given by El-Sayed (1977) and the new summary by Llano (1977).

The oceanic regions under discussion are characterized by seasonal fluctuation in the limits of the pack ice surrounding the continent, and by strong seasonal differences in light intensity. Water temperatures are more or less uniform, and differ only very slightly with season and depth. Primary production would at first sight appear to depend solely upon the seasonal alterations in light intensity, and to be directly correlated with it. However, this is not the whole story. Although it was previously assumed that phytoplankton photosynthesis was unlikely at the low light intensities beneath the ice, evidence is accruing from both Arctic and Antarctic that there is sufficient light at the layer where ice and water meet for fairly high production to take place. The layer is colonized by algae and plays a not unimportant role in the productivity of a cold oceanic region. The usual procedures for measuring photosynthesis in open water apparently give unsatisfactory results in this case.

The main beneficiary of the algae of the primary production in Antarctic regions is krill. This is a crustacean or rather a whole series of Euphausia species, the most important of which is E. superba. The animals live a planktonic existence and achieve a length of 4–6.5 cm. Repeated attempts to exploit krill for human consumption have been made in the recent past. In view of the extraordinary importance of the animals a very thorough investigation is essential, and this is now in progress. Surprisingly, the krill is far from being circumpolar in distribution. Instead, the animals concentrate in specific regions (Fig. 156) where they form large swarms suspended at irregular intervals in the immense body of water. It is now possible to locate them with the aid of modern echo-sounding devices. At the moment it is still unknown how long the animals require to achieve sexual maturity, whether they then produce several broods in succession, or whether they die after producing their first and only brood. In general, it is assumed that krill attain maturity after four years and then produce offspring for a whole summer, although according to another theory they mature after only two years and subsequently reproduce for a further two years. If the protein resources of the Antarctic are to be intelligently utilized and not simply over-exploited as was the case with the whales, it is imperative that the growth pattern of krill be very exactly elucidated. A realistic estimate of the possible yields available to Man can only be based on productivity and not on stocks.

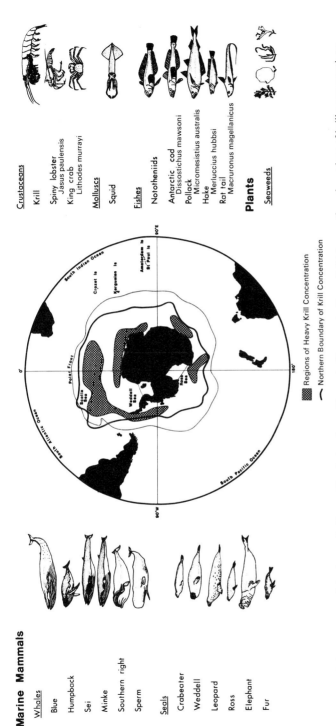

Fig. 156. Living resources of the Antarctic Ocean. *Hatched* regions of heavy krill concentrations. *Solid line* northern boundary of krill concentrations. *Dotted line* polar front. (El-Sayed, 1977)

It seems logical to assume that the numbers of krill must have undergone a change as a result of the decimation of the whale populations. This is of considerable relevance in view of the fact that it is the marine mammals and birds and not the fish that are the principal consumers of krill, poikilotherms being at a considerable disadvantage in the constant low temperatures of antarctic waters. Marine mammals and birds head the list in the exploitation of the live resources of the Antarctic. Table 27 shows the size of the various seal populations and their contribution to energy flow. It should be borne in mind that the stocks of sea elephants and fur seals were largely destroyed by Man, and that the populations are only now beginning to show signs of regeneration. In the case of the fur seals on South Georgia, however, this is proceeding relatively rapidly. Although in the 1930s survivors numbered only a few hundred, figures have now increased to about 350,000. However, approximately another 15 or 20 years of undisturbed multiplication will be necessary for complete restoration of the earlier stocks of several millions.

In the case of whales, the initial numbers used as a basis of comparison with present-day figures were themselves already influenced by Man (Table 27). The massive reductions (with the exception of the sperm whale which is the favoured victim of whale-hunters to-day) must have involved a significant drop in the consumption of krill. There are indications that not only the rapid increase of seal populations but also the earlier attainment of sexual maturity by Weddell seals nowadays is connected with the fact that food supplies are better, even though the majority of whales spend only a part of the year in the Antarctic. Only the sperm whale spends the entire year in antarctic waters, whereas all others breed in tropical latitudes, where they consume very little food. It is estimated that during the period of intensive feeding in antarctic waters the weight of a whale increases by about 50%, and that its reserves are depleted during the winter in tropical seas.

Birds, too, chiefly the penguins which live mainly on krill, appear to have profited from the decrease in the whale stocks; their colonies have apparently expanded as a result (Table 27).

However, certain reservations have to be made with respect to the values shown in the tables, mainly due to uncertainty regarding the real food requirements of the animals concerned. It is usually assumed, as in Table 27, that the basic metabolism of the animals is twice that of comparable terrestrial homeotherms (for mammals: kcal per day $= 2 \times (70 \times KGW^{0.75})$; for birds: kcal per day $= 2 \times (89\, KGW^{0.73})$), to which has then to be added the energy required for daily activities such as catching prey, display etc. However, data gathered in zoological gardens reveal a completely different situation. Here, seals and dolphins are fed a quantity of fish that, assuming a digestive efficiency of 80%, corresponds to little more than the normal basic metabolic value would require. If these feeding data are correct the estimates concerning plankton and nekton requirements of marine mammals would have to be thoroughly revised (see p. 126, Table 7, Fig. 96).

It therefore appears that in antarctic waters a single primary consumer is exploited almost equally by the rest of the consumers. The coexistence of so many different species reliant upon one and the same food source is a remarkable ecological phenomenon, with the additional unexpected situation that, as far as we know, fish play no significant role in the ecosystem of antarctic waters. El-Sayed

Table 27a–c. A preliminary estimate of the role of mammals and birds in the energy flow of the Antarctic Ocean. (El-Sayed, 1978)

Table 27a

Species	Stock (thousands)	Mean weight (kg)	Population biomass[a]	Annual food consumption[a]			
				Total	Krill	Squids	Fishes
Elephant	600	500	300	6,000	—	3,000[b]	3,000[b]
Leopard	220	272	60	1,403	519	112	182
Weddell	730	246	180	4,211	—	463	2,232
Crabeater	15,000	193	3,000	67,245	63,210	1,345	2,017
Ross	220	173	38	892	80	571	196
Fur	350	50	17.5	410	205	102	102
Total[c]	17,000	—	3,600	80,000	64,000	6,000	8,000

[a] Thousands of metric tons
[b] Assumed proportions in absence of pelagic data
[c] Total rounded

Table 27b

Species	Stock (thousands)	Mean weight (metric tons)	Mean Antarctic biomass[a]	Food consumption in Antarctic[a]			Biomass loss outside Antarctic[a]
				Krill	Squids	Fishes	
Initial stocks							
Baleen whales							
Fin	400	50	20,000	81,480	840	1,680	
Blue	200	88	17,600	71,702	740	1,478	
Sei	75[b]	18.5	1,387	5,651	58	116	
Humpback	100	27	2,700	11,000	113	227	
Minke	200	7	1,400	10,000[d]	204	409	
Total	975		43,000	180,000	2,000	4,000	18,000
Sperm Whale	85[c]	30	2,600		10,200	500	600
Present stocks							
Baleen whales							
Fin	84	48	4,032	16,426	169	339	
Blue	10	83	830	3,381	35	70	
Sei	40.5[b]	17.5	709	2,888	30	60	
Humpback	3	26.5	79	322	3	7	
Minke	200	7	1,400	10,000[d]	204	409	
Total[e]	338		7,000	33,000	440	880	2,500
Sperm Whale	43[c]	27	1,200		4,600	250	250

[a] Thousands of metric tons
[b] Half total Sei Whale stock assumed to feed south of the Antarctic Convergence
[c] One third total "exploitable" Sperm Whales assumed to feed in the Antarctic
[d] Minke Whale assumed to feed for six months in Antarctic waters at 4 per cent body weight/day
[e] Totals rounded

Table 27c

	Stocks (thousands)	Biomass[a]	Food consumption in Antarctic and sub-Antarctic[a]
Spheniscidae	121,770	520.7	33,173
Diomedeidae	12,630	44.45	2,867
Procellariidae	39,354	10.17	2,131
Pelecanoididae Hydrobatidae	11,000	1.3	427
Phalcrocoracidae Stercoraciidae Laridae	2,835	0.56	65
Total[b]	188,000	580	39,000

[a] Thousands of metric tons
[b] Totals rounded

(1977) gives a list of species that may be of commercial interest in the future. Some of these are already being exploited, but on the whole prospects seem to be poor. The fish are for the most part small, rarely occur in shoals, and are chiefly found in the immediate vicinity of the coasts, all of which are factors unfavourable to modern methods of exploitation. The situation with regard to squids is still unclear. Reasonable numbers apparently occur pelagically, and are at present being exploited to some degree by the Japan fishery (and by some of the endotherms, such as the king penguin).

The underlying reason for the unique composition of the nutritional chain is probably to be seen in the very low temperatures of antarctic waters. Primary production is relatively high, as is shown in Fig. 25, but cannot be utilized by the marine endotherms because the particle size is too small. In order to be able to take up particles of this size a homeotherm would require a very much thicker suspension, or it would have to be a very small animal. But since small warm-blooded animals cannot exist in this permanently cold environment the primary consumers are poikilothermic organisms (Krill). These can be consumed by the large warm-blooded animals since the particles are large enough and, at the prevailing temperatures, the large homeotherms are at a very clear advantage as compared with the poikilothermic fish. Although the data on growth rate of fish in antarctic regions (i.e., at constant temperatures of about $-1.5°--2.0°$ C) are extremely unsatisfactory, the little data that we have indicate that all species take 10–20 years to achieve maturity. In fact, growth is still comparatively fast, but the animals have forfeited a large part of their mobility, tolerance to temperature fluctuations, and reproductive potential. They expend a substantial portion of their energy on producing the glycoproteins that serve as antifreeze agents.

With baleen whales as end-consumers, krill as primary consumer and planktonic algae as producers, the nutritional chain in the Antarctic Ocean is not only the shortest known in the world, it also leads to the largest animals.

The islands in the Antarctic Ocean, South Georgia, Marion, Macquarie, Kerguelen, are characteristically termed the "subantarctic islands". We have to take care not to lose sight of the fact that practically the entire Antarctic Ocean and these islands are outside the Polar Circle, and thus beyond the limit we employed as a definition in the case of the Arctic. As already seen from Fig. 3 the climate on the subantarctic islands is extremely stable, without cold winters or warm summers, and with barely detectable diurnal fluctuations in temperature, which is very constant and low throughout the year. Conditions are thus completely different from those obtaining in the Arctic. Photosynthesis can go on throughout the entire year and primary production appears to be correspondingly high. Of the animals, it is the warm-blooded marine forms such as various species of penguin, Procellaridae, albatrosses, fur seals and sea elephants that play by far the most important role. The resultant eutrophication of the islands, combined with the practically uninterrupted growth season, leads to a luxuriant vegetation. The antarctic ectothermic fauna is thus also totally different from that of the Arctic, and in contrast to the situation in the latter, herbivorous beetles apparently dominate (e.g., Smith, 1977). On the whole, the ectothermic fauna is very small. The true terrestrial vertebrate fauna is exceptionally poor, the sole representatives being a freshwater duck and a pipit (Anthus), and these only on very large islands.

Obviously the Antarctic is no place for warm-blooded terrestrial predators, in striking contrast to the situation in the Arctic, where wolves, weasels, polar foxes and bears are found even in the northernmost latitudes. In some arctic regions such as Spitsbergen, solitary polar bears can exist on the ice, more or less independent of land. There are many reports of polar foxes covering large distances across the ice, and the species is even believed to have reached Iceland and Greenland in this way: after its extermination on Bear Island recolonization probably took place across the winter ice. The absence of all such predatory forms from the Antarctic has been a decisive factor in the evolution of its birds and mammals. Whereas arctic birds breed mainly on unassailable vertical cliffs, safe from marauding polar foxes, bears and wolves, or on offshore islands that are too small to offer their enemies a permanent abode, the large antarctic penguin colonies are situated on the flat beaches of islands and mainland. The evolution of the penguin's inability to fly probably resulted from the lack of pressure from enemies. Prior to its becoming extinct, the flightless great auk of the Northern Hemisphere, roughly comparable to the king penguin, nested exclusively on small offshore islands where terrestrial predators were unable to exist for longer periods of time. The arctic seals (hooded seal, harp and ringed seal for example) bear a single pup with the famous snowy white coat that renders it practically undetectable in its icy surroundings. The young of the antarctic seals, on the other hand, that have no enemies, enter the world with a darker coat. Again, the antarctic seals leave their excrements and even placentas during the breeding season around the holes in the ice where they rest. Arctic seals, however, apparently defaecate in the water and either eat the afterbirth or push it into the water so as not to leave any characteristic odour that might attract a hungry polar bear.

In the antarctic and subantarctic islands the function of the terrestrial predators of the Arctic is taken over by the various forms of the great skua, and by the sheath bill (Chionis). We are indebted to Müller-Schwarze (in Llano, 1977) for an evaluation of the role played by the great skua. An adult requires about 500 g of penguin egg or flesh per day, which is the equivalent of 7 eggs or a comparable quantity of penguin chick. The following calculation has been made for a large colony of Adelie penguins on the antarctic continent, with 6 territorial pairs of great skua and 1600 penguins pairs:

For the 1.81 eggs laid per penguin pair, 1.08 chicks per pair leave the colony in a fully grown state. The loss is almost entirely due to the skuas, which means that each skua consumes 97 eggs or young animals. In actual fact, however, it can be assumed that an adult skua, in order to cover its entire food requirements at the expense of the penguin colony, would require 296 penguin eggs or chicks in this period of time. Even then, this completely ignores the feeding of the young skuas. This means that the skuas living in the vicinity of a penguin colony do far less damage than would be expected. A possible explanation is provided by the territorial behaviour of the birds. The penguin colony is split up between a few skua pairs, and these permit no other member of their species to hunt in their territory. This is of great advantage to the penguins, because if the skuas were not to defend their territory from other members of their species the penguins would very quickly disappear altogether. A similar state of affairs is reported from other penguin colonies where apparently a large number of the skuas, i.e., those that have no territory of their own within the penguin colony, hunt their food far from their nesting places. The very rough outline of the Antarctic given at the outset of this case study is inadequate for a consideration of its continental terrestrial fauna, and further subdivision becomes necessary. Where the ocean cuts deeply into the antarctic continent at the Ross and Weddell shelves there are ice-free areas in summer up to a southern latitude of 76°, and this is the southernmost latitude at which life on land is possible and demonstrable. At the other extreme, the Palmer peninsula extends far north of the true antarctic continent to a latitude of about 63° south.

For this reason the peninsula is often considered separately from the antarctic continent, as for example in ornithological lists, and in fact its plant and animal life do differ greatly from those of the actual continent. The only two vascular plants found today in the Antarctic (Colobanthus crassifolius and Deschampsia antarctica) and the marine chironomid Belgica antarctica are found solely here. The only place where a freshwater chironomid (Parochlus steineni) is encountered is in the northernmost part of the Palmer peninsula. It is the only free-living non-marine winged insect of the Antarctic (with the sole exception of forms parasitizing on homeotherms and in their nests, Table 28). Although even on the Palmer peninsula the growth of Collembola is very slow (so that Tilbrook, in Sayed, 1977, in distinguishing size classes was unable to detect any growth over one year) the energy flow through a population is apparently fairly high. On Signy Island 60,000–61,000 Collembola of the species Cryptopygus antarcticus can be counted per m^2 of ground. This corresponds to a biomass of almost 800 mg per m^2. The Collembola consume a total of 1.2 ml oxygen per m^2 and year, or the equivalent of 5.8 cal. Production of Collembola substance annually is put at 1.2 mg or 2.6 cal per m^2 and year, and the food consumed is estimated to be 2–4 mg per m^2 and year (Tilbrook in Llano, 1977). The 14 bird species breeding on the peninsula include the giant fulmar (Macronectes giganteus), the cape pigeon (Daption capensis), the southern fulmar (Fulmarus glacialoides), the snow petrel (Pagodroma nivea), Wilson's petrel (Oceanites oceanicus), the cormorant (Phalacrocorax atriceps), the sheath bill (Chionis alba), the great skua (Stercorarius skua), the black-backed gull (Larus

Table 28. Insects parasitizing on warm-blooded organisms of the Antarctic. (Murray in Cheng, 1976)

Endotherm	Insect
Mirounga leonina	Lepidophthirius macrorhini Enderlein
Hydrurga leptonyx	Antarctophthirius ogmorhini Enderlein
Leptonychotes weddeli	A. ogmorhini Enderlein
Lobodon carcinophagus	A. lobodontis Enderlein
Ommatophora rossi	A. mawsoni Harrison
Arctocephalus pusillus	Antarctophthirius callorhini (?)
Aptenodytes, Eudyptes, Eudyptula, Megadyptes, Pygoscelis Spheniscus	Austrogonioides, Nesiotinus (Mallophaga)
Fulmarus, Pachyptila, Pagodroma, Thalassoica	Many mallophage genera, including Ancistrona, Austromenopon, Longimenopon, Bedfordiella, Docophoroides

dominicanus), and the tern (Sterna vittata). In addition, there are the penguins, although it is impossible to decide which species extend into the true Antarctic and which are confined to the islands. The emperor penguin (Aptenodytes forstere) is encountered only at the base of the peninsula, close to the actual continent, but the Adelie penguin (Pygoscelis adeliae), the Gentoo penguin (P. papua) and the ringed penguin (P. antarctica) are frequently encountered. On the continent itself there are no vascular plants and no pterygotous insects, and the penguins are reduced to the two species Aptenodytes forstere and Pygoscelis adeliae. Other birds nesting here include the great skua (Stercorarius skua) and the giant fulmars (Macronectes giganteus) and some further species (Thalassoica antarctica, Fulmarus glacialoides, Daption capensis, Pagodroma nivea, Ptachyptila desolata and Oceanites oceanicus). Such species are of course confined to places where either ground or bedrock are exposed. The emperor penguin is the only species nesting directly on the ice. Poikilothermic animals found here include Collembola and mites, and Janetschek (1963) reported finding mites even at 78°S, where they feed on the lichens by piercing them with their sharp mouth parts. Of the mites it is the Trombidiformes that penetrate farthest into the Antarctic, and of the Collembola the Podurids. Their occurrence in nunataks is only rendered possible by the presence of lichens, and they and bdelloid rotifers are found here only in places that receive more or less regular sunshine. Janetschek distinguishes between three different systems. The most hospitable of these is a moss system (bryosystem), characterized solely by moss harbouring mites and Collembola. A contrasting system consists of pebbles and rubble (chalikosystem). It is devoid of plant life and frequented by a few arthropods that feed on soil fungi. Between the two systems is a third and intermediary type which is a mosaic of the others. Such systems are mainly of interest to physiologists, since they provide valuable examples of organisms clearly indicating the cold limits to existence. The systems are of no significance for the antarctic ecosystem as a whole, for which the colonies of aquatic birds mentioned previously are of far greater importance.

Antarctic freshwater systems are surprisingly rich, considering the prevailing temperatures. It was here that the discovery was first made that lakes with a permanent covering of ice can act as a trap for solar energy, and can therefore support planktonic and benthic plant and animal life (blue algae, a few green algae, rotifers; Hoare et al., 1964, 1965).

At least 11 aquatic crustaceans belonging to systematic groups that are very reminiscent of High-Arctic conditions occur in the less exposed lakes of the Palmer peninsula (Branchinecta species, Copepoda, Ostracoda and Cladocera). It is an interesting point that here many mosses are found in freshwater, thus avoiding the drying out that represents such a great danger to plants in polar environments. Certain moss species flourish in antarctic lakes down to a depth of 35 m and more (Heywood in Llano, 1977).

Clearly, it is a much easier matter to establish subdivisions in the Antarctic than in the Arctic. Excluding the Palmer peninsula, the antarctic continent can therefore be termed "High Antarctic"; there are no vascular plants and no winged insects, but merely mosses and lichens, mites and Collembola. Although beetles have been reported on the South Shetland and South Orkney islands it is not certain whether they might not have been imported by Man, nor to what extent they are connected with human activities on these islands (Balsour-Brown and Tilbrook in El-Sayed, 1966). Terrestrial birds and mammals are also absent, and every High-Antarctic bird or mammal is immediately dependent upon the sea for its existence. Since the temperature of the surrounding seas is everywhere below zero, the area can therefore only harbour cold-adapted forms. The continent is surrounded by islands, to which the Palmer peninsula can also be reckoned. Here, in the southern summer, temperatures rise above freezing point, which is a rare occurrence on the main continent. Terrestrial birds such as the sheath bill (Chionis) are encountered here for the first time, as well as a few vascular plants and winged insects. All of these islands lie within the ice regions, and are shut in by ice in winter. High-Antarctic birds and mammals, like the emperor penguin, the Ross seal and the Weddell seal are missing, but are replaced by the sea leopard, the crab-eater, the antarctic cormorant and the black-backed gull. Farther north are the subantarctic islands, with their extremely even climate, where summer and winter are almost identical, although temperatures are still constantly low. These islands lie beyond the ice region and thus temperatures at sea level are almost always above zero. Fur seals, sea elephants, king penguins and the first real land birds (pipit and freshwater duck) are encountered here. Insects are represented by a wealth of mainly flightless species, some of which are herbivorous beetles.

The Antarctic is of such interest today because of its potentialities for mankind as an unexploited source of nutritional protein. The bitter experience connected with earlier human efforts in this region has prompted the organization of extensive international programmes to ensure a thorough investigation of the extent of the resources available and to analyse possible ways of using them. Undoubtedly vast quantities of energy-rich protein for human consumption could be harvested from antarctic waters over unlimited periods of time if over-exploitation were avoided.

The current intensity with which research is being carried on to this end in the Antarctic renders it impossible to offer a survey, comparable to that attempted for the Arctic, that would not be obsolete before it could be printed.

References

Andersen F Søgaard (1946) East greenland lakes as habitats for chironomid larvae. Medd Groenl 100:5-65

Andersen HT (ed) (1969) Biology of marine mammals. 511 pp. Academic Press, London New York

Andersson G, Gerell R, Källander H, Larsson T, (1967) The density of birds in some forest habitats in store Sjöfallet National Park, Swedish Lapland. Acta Univ Lund II 17:3-11

Andersson NA, Müller K (1978) Der Tagesrhythmus des Stares Sturnus vulgaris und anderer Singvögel in Abisko, Nordschweden. Ornis Scand 9:40-45

Andrews RV, Ryan K, Strohbein R, Ryan-Kline M (1975) Physiological and demographic profiles of brown lemmings during their cycle of abundance. Physiol Zool 48:64-83

Anonym (1970) Proceedings of the Conference on Productivity and Conservation in Northern Circumpolar Lands (Edmonton, Alberta, 1969) IUCN Publications N S, No 16, 344 pp. Morges, Switzerland

Anonym (1974) Hardangervidda. Norges offentlige Utredninger Miljöverndepartementet, 3 vek. Universitetsförlaget Oslo

Arntz WE (1974) Die Nahrung juveniler Dorsche (Gadus morhua L.) in der Kieler Bucht. Ber Dtsch Wiss Komm Meeresforsch 23:97-120

Aschoff J, Gwinner E, Kureck A (1970) Diel rhythms of chaffinches Fringilla coelebs L., tree shrews Tupaia glis L. and hamsters Mesocricetus auratus L. as a function of season at the Arctic Circle. Oikos suppl 13:91-100

Aschoff J, Daan S, Figala J, Müller K (1972) Precision of entrained circadian activity rhythms under natural photoperiodic conditions. Naturwissenschaften 59:276-277

Bagge P (1968) Ecological studies on the fauna of subarctic waters in Finish Lapland. Ann Univ Rurku Ser A 12 40:32-40

Balfour-Browne J, Tilbrook PJ (1966) Coleoptera collected in the South Orkney and South Shetland Islands. Br Antarct Surv Bull 9:41-43

Bardin W (1970) Oasen der Antarktika. Bild Wiss 1:47-53

Batzli GO (1975) The role of small mammals in arctic ecosystems In: Golley, Petrusewicz, Ryszkowski (eds) Small mammals: their productivity and population dynamics, University Press, Cambridge pp 223-242.

Beck L (1972) Zur Tagesperiodik der Laufaktivität von Admetus pumilio C. Koch (Arach Amblypygi) aus dem neotropischen Regenwald II. Oecologia 9:65-102

Behrisch H (1973) Molecular mechanisms of temperature adaption in arctic ecotherms and heterotherms. In: Wieser (ed) Effects of temperature on ectothermic organisms, p 298. Springer, Berlin Heidelberg New York

Belopolski LO (1957) Die Ökologie der kolonialen Seevögel der Barents-See, pp 3-458. Moskau-Leningrad (in Russian)

Bengtson S-A (1976) Effect of bird predation on lumbricid populations. Oikos 27:9-12

Bengtson S-A, Fjellberg A, Solhöy T (1974) Abundance of Tundra Arthropods in Spitsbergen. Entomol Scand 5:137-142

Bergmann HH, Klaus S, Müller F, Wieser J (1978) Das Haselhuhn, S. 196. Neue Brehmbücherei, Wittenberg-Lutherstadt

Bertsch A (1977) In: Trockenheit und Kälte. Anpassung an extreme Lebensbedingungen, S. 143. Dynamische Biologie, Bd 6. Maier, Ravensburg

Bie de Steven (1977) Survivorship in the Svalbard reindeer (Rangifer tarandus platyrhynchus Vrolik) on Edgeøya, pp 249-270. Svalbard, Norsk Polarinstitut Årbok 1976, 1977

Billings WD (1973) Arctic and Alpine vegetations: Similarities, differences, and susceptibility to disturbance. Bioscience 23:697-704

Bliss LC (ed) (1977) Truelove Lowland, Devon Island, Canada - a high arctic ecosystem, 714 pp. University of Alberta Press, Edmonton

Bliss LC, Wielgolaski FE (1973) Primary production and production processes, tundra biome, pp 3-256. Tundra Biome Steering Committee, Stockholm

Blotzheim U, Glutz N (1973) Handbuch der Vögel Mitteleuropas, Bd 1-5. Akademische Verlagsgesellschaft, Wiesbaden

Bovet J, Oertli EF (1974) Free-running circadian activity rhythms in free-living beaver (Castor canadensis). J Comp Physiol 92:1-10

Brown J (1975) Ecological investigations of the tundra biome in the Prudhoe Bay Region, Alaska. Biol Pap Univ Alaska Spec Rep 3-215

Brown RGB (1963) The behaviour of the willow warbler Phylloscopus trochilus in continuous daylight. Ibis 105:63-75

Brown RGB, Nettleship D, Germain P, Tull CE, Davis T (1976) Atlas of Eastern Canadian seabirds, 220 pp. Information Canada, Ottawa K1A OS9

Brown RJE, Johnston GH (1964) Permafrost und verwandte Ingenieurprobleme. Endeavour 66-72

Brzoska W (1976) Produktivität und Energiegehalte von Gefäßpflanzen im Adventdalen (Spitzbergen). Oecologia (Berlin) 22:387-398

Buinitsky VKh (1977) Organic life in sea ice. In: Dunbar (ed) Polar oceans, pp 301-317. Arctic Inst America, Calgary

Burt W, Grossenheider RP (1964) A field guide to the mammals, 284 pp. Peterson Field Guide Ser

Cheng L (1976) Marine insects, 581 pp. North Holland Publ Co, Amsterdam Oxford

Cody ML (1974) Bird communities. Monogr Popul Biol 7:318

Corbet PS (1966) Diel periodicities of weather factors near the ground in a high arctic locality. Hazen Camp, Ellesmere Island, N.W.T. Rep No 16:29

Corbet PS (1967) Further observations on diel periodicities of weather factors near the ground at Hazen Camp, Ellesmere Island N.W.T. D Phys R (G) Hazen 31 (Report)

Crowther AGD (1959) Ornithological results of the Cambridge Vannøy expedition 1958 A. 48 Hours observation on mealy redpoll nest. Astarte 17:2-13

Croxall JP, Kirkwood ED (1979) The distribution of penguins on the antarctic peninsula and islands of the Scotia Sea, pp 1-186. Br Antarct Surv Madingley Road, Cambridge CB3 OET

Cullen JM (1954) The diurnal rhythm of birds in the arctic summer. Ibis 96:31-46

Daan N (1975) Consumption and production in north sea cod, Gadus morhua: an assessment of the ecological status of the stock. Neth J Sea Res 9:1 24-55

Daan S, Aschoff J (1975) Circadian rhythms of locomotor activity in captive birds and mammals: their variations with season and latitude. Oecologia 18:269-316

Dahl C (1973) 14. Trichoceridae (Dipt.) of Spitzbergen Notes on the arthropod fauna of Spitzbergen III. Ann Entomol Fenn 39:49-59

Dauphiné TC Jr (1976) Biology of the Kaminuriak population of barren-ground caribou, part 4, Growth, reproduction and energy reserves. Can Wildl Serv Rep Ser N 38:71

Demmelmeyer H (1974) Endogene Veränderungen der Empfindlichkeit einiger Singvögel gegenüber schwachen tagesperiodischen Zeitgebern und die Beeinflussung dieser Empfindlichkeit durch Sexualhormone Dissertation 1-31

Demmelmeyer H, Haarhaus D (1972) Die Lichtqualität als Zeitgeber für Zebrafinken (Taeniopygia guttata) J Comp Physiol 25-29, 70:

Dietrich G, Kalle K, Krauss W, Siedler G (1975) Allgemeine Meereskunde, S 475, 3 Aufl. Bornträger, Berlin

Dircksen R (1962) Vogelvolk auf weiter Reise, S 191, Bertelsmann, Gütersloh

Downes JA Adaptations of insects in the Arctic. Ann Rev Entomol 10:257-274

Dunbar MJ (ed) (1977) Polar oceans, 681 pp. Arctic Inst North America, Calgary

Ekmann S (1953) Zoogeography of the sea, 417 pp. London

El-Sayed Z (1977) Biological investigations of marine antarctic systems and stocks (biomass), pp 1-79. Scar and Scor, Scott Polar Research Institute, Cambridge England

Enemar A, Hanson SA, Sjöstrand B (1965) The Composition of the bird fauna in two consecutive breeding seasons in the forests of the ammarnäs area, swedish lapland. Acta Univ Lund II 5:3-11

Enright J (1976) Climate and population regulation. The biogeographer's dilemma. Oecologia 24:295-310

Erkinaro E (1969a) Free running circadian rhythm in wood mouse (Apodemus flavicollis Melch.) under natural light-dark-cycle. Experientia 25:649

Erkinaro E (1969b) Der Verlauf desynchronisierter, circadianer Periodik einer Waldmaus (Apodemus flavicollis) in Nordfinnland. Z Vergl Phys 64:407-410

Erkinaro E (1970) Wirkung von Tageslänge und Dämmerung auf die Phasenlage der 24-h Periodik der Waldmaus Apodemus flavicollis Melch. im Naturtag. Oikos Suppl 13:101-107

Erskine AJ (1980) A preliminary catalogue of bird census plot studies in Canada, part 4. Progress Notes Nr. 112, Canadian Wildlife Service

Eurola S (1971) The driftwoods of the Arctic Ocean. Rep Kevo Subarct Res Stat 7:74-80

Eurola S (1972) Germination of seeds collected in Spitsbergen. Ann Bot Fenn 9:149-159

Eurola S, Hakala AVK (1977) The bird cliff vegetation of Svalbard. Aquilo Ser Bot 15:1-18

Ferenz H-J (1975) Anpassung von Pterostichus nigrita F. (Col Carab) an subarktische Bedingungen. Oecologia (Berlin) 19:49-57

Feyling-Hansen RW (1953) Balanus balanoides in Spitzbergen. Norsk Polarinst, Skrifter Nr 98

Fisher J (1952) The fulmar, 496 pp. Collins, London

Fisher J (1954) Birds as animals I. A history of birds, 205 pp. Oxford

Fisher J, Lockley RM (1954) Sea-birds. An introduction to the natural history of the sea-birds of the North Atlantic. Collins, London

Fisher J, Rosin S (1978) Einfluß von Licht und Temperatur auf die Schlüpfaktivität von Chironomus nuditarsis. Rev Suisse Zool 75:538-549

Franz J (1943) Über Ernährung und Tagesrhythmus einiger Vögel im arktischen Winter. J Ornithol 91:154-165

Franz J (1948) Jahres- und Tagesrhythmus einiger Vögel in Nordfinnland. Z Tierpsychol 6:309-329

Gardasson A (ed) (1975) Votlendi. Rit Landverndar. 4:238

Gardasson A (1979) Waterfowl populations of Lake Myvatn and recent changes in numbers and food habits. Oikos 32:250-270

Gardasson A, Sigurdsson JB (1972) Skyrsla um Rannsoknir a Heidagaes i Thorsaverum sumarid 1971. Prog Rep 100

Gardasson A, Sigurdsson JB (1974) Skyrsla um Rannsoknir i Thorsaverum 1972. Prog Rep 100

Gerlach SA (1965a) Tierwanderungen. Handb Biol 5: Heft 14-19, 413-472

Gerlach SA (1965b) Freilebende Meeresnematoden aus der Gezeitenzone von Spitzbergen. Veröff Inst Meeresforsch Bremerhaven 9:109-172

Gewalt W (1976) Der Weißwal Delphinapterus leucas, 232. Neue Brehm Bücherei, Wittenberg

Gill D (1974) Significance of springbreakup to the bioclimate of the Mackenzie River delta, In: Reed JC, Sater JE (eds) The coast and shelf of the Beaufort Sea, 543 pp. Arctic Institute of North America; Arlington

Gjaerevoll O, Jørgensen R (1953) Fjällflora. PA Norstedt Söners, Stockholm

Golikov AN, Averincev VG (1977) Distribution patterns of benthic and ice biocoenoses in the high latitudes of the Polar Basin and their part in the biological structure of the World Ocean. In: Dunbar (ed), Polar oceans, pp 331-364, Arctic Inst America, Calgary

Gossow H (1974) Natural mortality pattern in the Spitsbergen. reindeer, pp 103-106. 11th Int Congr Game Biol, Stockholm

Gossow H, Thorbjørnsen A (1974) Air and ground survey of reindeer in Nordenskiöld Land and Sabine Land, Spitsbergen, pp 83-88. Norsk Polarinst Årbok, Oslo

Grainger EH (1953) On the age, growth, migration, reproductive potential and feeding habits of the Arctic char (Salvelinus alpinus) of Frobisher Bay. Baffin Island. J Fish Res Board Can 10:326-370

Gray DR (1970) The killing of a bull muskox by a single wolf. Arctic 23:197-199

Green RH (1973) Growth and mortality in an Arctic intertidal population of Macoma balthica. J Fish Res Board Can 30:1345-1348

Gudmundsson F (1960) Some reflections on ptarmigan cycles in Iceland, pp 259-265. Proc XIIth Int Ornithol Congr, Helsinki

Gurjanoca EF (1968) The influence of water movements upon the special composition and distribution of the marine fauna and flora throughout the arctic and North Pacific intertidal zones. Sarsia 34:83-94

Haarhaus G (1968) Zum Tagesrhythmus des Staren (Sturnus vulgaris) und der Schneeammer (Plectrophenax nivalis). Oecologia 1:176-218

Hackmann W, Nyholm E (1968) Notes on the arthropod fauna of Spitsbergen II 9. Mallophaga from Spitsbergen and Bear Island. Ann Entomol Fenn 34:75-104
Haftorn S (1971) Norges Fugler, 862 pp. Oslo
Haftorn S (1972) Hypothermia of tits in the Arctic winter. Ornis Scand 3:153-166
Hall AB (1966) The breeding birds of an east Greenland valley, 1962. Dan Ornithol Foren Tidsskr 60:175-185
Hall AB, Waddingham RN (1966) The breeding birds of Ørsteds Dal, East Greenland 1963. Dan Ornithol Foren Tidsskr 60:186-197
Harington RC (1968) Denning habits of the bear (Ursus maritimus Phipps.). Can Wildl Serv Rep Ser 5:30
Haukioja E, Hakala T (1975) Herbivore cycles and periodic outbreaks. Formulation of a general hypothesis. Rep Kevo Subarct Res Stu 12:1-9
Haukioja E, Niemelä P (1977) Retarded growth of geometrid larvae after mechanical damage to leaves of its host tree. Ann Zool Fenn 14:48-52
Hedgpeth JW (1957) Treatise on marine ecology and paleoecology, vol I. Ecology Memoir 67. Geol Soc Am Mem 67:1-1296
Heinrich B (1973) Mechanism of insect thermoregulation. In: Wieser W (ed) Effects of temperature on ectothermic organisms, pp 139-150. Springer, Berlin Heidelberg New York
Hempel G, Hempel I (1955) Über die tägliche Verteilung der Laufaktivität bei Käfern des Hohen Nordens. Naturwissenschaften 42:77-78
Hempel G (1977a) Fischerei in marinen Ökosystemen. Verh Dtsch Zool Ges Erlangen 67-86, Stuttgart New York
Hempel G (1977b) Biologische Probleme der Befischung mariner Ökosysteme. Naturwissenschaften 64:200-206
Hempel G (1979a) Meeresfischerei als ökologisches Problem. Rheinisch Westfäl Akad Wiss (Vortr) 283:8-48
Hempel G (1979b) Fischereiregionen des Weltmeeres - Produktion und Nutzung. Geogr Rundschau 31 12:492-497
Hempel G (1980) Das Leben in Eis und Schnee. Bild Wiss 1:37-48
Herre W (1956) In: Ziemsen A (Hrsg) Rentiere. S 3-47, Heft 180 A Ziemsen, Neue Brehm Bücherei, Wittenberg Lutherstadt
Higgins LG, Riley ND (1975) Die Tagfalter Europas und Nordwestafrikas. Hamburg
Hinz W (1976) Zur Ökologie der Tundra Zentralspitsbergens. Nor Polarinst Skr 163:1-47
Hoare RA, Popplewell KB, House DA, Henderson RA, Prebble WM, Wilson AT (1964) Lake Bonney, Taylor Valley, Antarctica: a natural solar energy trap. Nature (London) 202:886-888
Hochachka PW, Somero GN (1973) Strategies of biochemical adaptation, pp 1-358. Saunders, Philadelphia
Hocking B (1968) Insect-flower associations in the Arctic with special reference to nectar. Oikos 19:359-388
Høeg OA (1932) Blütenbiologische Beobachtungen aus Spitzbergen. Nor Svalbard-Og Ishavs-Undersøkelser Medd 16:3-22
Hoffmann K (1959a) Über den Tagesrhythmus der Singvögel im arktischen Sommer. J Ornithol 100:84-89
Hoffmann K (1959b) Die Richtungsorientierung von Staren unter der Mitternachtssonne. Z Vergl Physiol 41:471-480
Hoffmann K-H (1974) Wirkung von konstanten und tagesperiodisch alternierenden Temperaturen auf Lebensdauer, Nahrungsverwertung und Fertilität adulter Gryllus bimaculatus. Oecologia (Berlin) 17:39-54
Hoffmann K-H (1976a) Organic body constituents of Protophormia terrae-novae (Dipt) from Spitsbergen compared with Flies from a Laboratory stock. Oecologia 23:13-16
Hoffmann K-H (1976b) Catalytic efficiency and structural properties of invertebrate muscle pyruvate kinases: correlation with body temperature and oxygen consumption rates. J Comp Physiol 110:185-195
Hoffmann K-H (1977) The regulatory role of the muscle pyruvate kinase in carbohydrate metabolism of invertebrates: a comparative study in catalytic properties of enzymes isolated from tubifex tubifex (oligochaeta) and tenebrio molitor (coleoptera). Physiol Zool V 50:No 2, 142-155
Hoffmann K-H (1978a) Thermoregulation bei Insekten. Biol Unserer Zeit 8 Jahrg 1:17-26

Hoffmann K-H (1978b) Anpassung in der Konformation von Proteinmolekülen an die Temperaturökologie einer Art. Verh Dtsch Zool Ges Konstanz, S 214. Fischer, Stuttgart New York

Hoffmann K-H, Friedrich E (1977) Temperatur-induzierte Differenzierung des Enzym-Aktivitätsmusters der Mittelmeer-Feldgrille, Gryllus bimaculatus (Orthoptera: Gryllidae). Ent Germ 3(4):289-297

Hoffmann K-H, Marstatt H (1977) The influence of temperature on catalytic efficiency of pyruvate kinase of crickets (Orthoptera: Gryllidae). J Therm Biol 2:203-207

Hoffmann K-H, Dachlauer G, Glöckner H (1977) Temperaturinduzierte Veränderungen der Metabolitkonzentrationen im Kohlenhydrat-, Protein- und Lipidstoffwechsel von Gryllus bimaculatus (Orthopt.) während der Ovarentwicklung. Verh Dtsch Zool Ges 329

Holdgate MW (1970) Antarctic ecology. The Scientific Committee on Antarctic Research, Vol I-II. pp 3-604, 608-998. Academic Press, London New York

Holding AJ, Heal OW, MacLean SF, Flanagan PW (1973) Soil organisms and decomposition in tundra, pp 3-398. Tundra Biome Steering Committee, Stockholm

Irving L (1972) Arctic life of birds and mammals including man. Zoophysiol Ecol 2:192

Irving L, West G, Peyton C, Leonhard J (1967a) Winter feeding program of Alaska Willow Ptarmigan shown by crop contents. Condor 69:69-77

Irving L, West G, Peyton C, Paneak S (1967b) Migration of Willow Ptarmigan in Arctic Alaska. Arctic 20:77-85

Ives JD, Barry RG (1974) Arctic and alpine environments 999 pp. Methuen, London, and Harper & Row, New York

Jakimchuk RD, McCourt KM (1975) Distribution and movements of the porcupine caribou herd in the Northern Yukon. In: Luick JR (ed) Proc Ist Int Reindeer/Caribou-Symp Fairbanks University of Alaska

Janetschek H (1963) Zur Biologie von Antarktika. Ber Naturwiss Med Ver Innsbruck 53:235-246

Järvinen O, Väisänen RA (1976) Species diversity of Finnish birds, II: Biotopes at the transition between taiga and tundra. Acta Zool Fenn 145:1-35

Johnson SR, West GC (1975) Growth and development of heat regulation in nestlings, and metabolism of adult common and thick-billed murres. Ornis Scand 6:109-115

Johnstone GW (1974) Field characters and behaviour at sea of giant petrels in relation to their oceanic distribution, part 4. EMU 74:209-218

Johnstone GW, Kerry KR (1976) Proc 16th Int Ornithol Congr, pp 725-738. Aust Acad Sci PO Box 783, Canberra City, ACT 1601

Jònasson PM, Adalsteinsson H, Hunding C, Lindegaard C, Olafson J (1977) Limnology of Iceland. Folia Limnol Scand 17:111-123

Jònasson PM (ed) (1979) Ecology of eutrophic, subarctic Lake Myvatn and River Laxa. Oikos 32:1-308

Jonkel C, Smith P, Stirling I, Kolenosky GB (1976) The present status of the polar bear in the James Bay and Belcher Islands area. Can Wildl Serv Occ Pap 26:42

Kaisila J (1967) Notes on the arthropod fauna of Spitsbergen I. 1. Travel report. Ann Entomol Fenn 33:13-64

Kaisila J (1973a) 15. The Lepidoptera of Spitsbergen. Ann Entomol Fenn 39:60-65

Kaisila J (1973b) 16. The Anoplura and Siphonaptera of Spitsbergen. Ann Entomol Fenn 39:63-66

Kalela O (1971) Seasonal differences in habitats of the Norwegian lemming, Lemmus lemmus (L.), in 1959 and 1960 at Kilpisjärvi, Finnish Lapland. Ann Acad Sci Fenn A IV Biol 178:2-22

Kangas E (1967) 5. Identification of the Coleoptera collected by the Finnish Spitsbergen expeditions. Ann Entomol Fenn 33:41-43

Kangas E (1973) 18. Über die Coleopteranfauna in Spitzbergen. Ann Entomol Fenn 39:68-70

Kanwisher JW (1955) Freezing in intertidal animals. Biol Bull 109:56-63

Kanwisher JW (1959) Histology and Metabolism of frozen intertidal Animals. Biol Bull 116:258-264

Karplus M (1952) Bird activity in the continuous daylight of the arctic summer. Ecology 33:129-134

Kartaschew NN (1960) In: Ziemsen A (Hrsg) Die Alkenvögel des Nordatlantiks. S 5-153. Neue Brehm-Bücherei, Wittenberg Lutherstadt

Kerbes RH (1973) The nesting population of lesser snow geese in the eastern Canadian Arctic. Can Wildl Serv Rep Ser 35:47

Kevan PG, Keith D McE (1970) Collembola as pollen feeders and flower visitors with observations from the high arctic. Quaest Entomol 6:311-426

Kevan PG. Shorthouse JD (1970) Behavioural thermoregulation by high Arctic butterflies. Arctic 23: 269-279

Kightley SPJ. Smith RIL (1976) The influence of Reindeer on the Vegetation of South Georgia: I. longterm effects of unrestricted grazing and the establishment of exclosure experiments in various plant communities. Br Antarct Surv Bull 44:57-76

Klein WH, Goldberg B (1973) Solar radiation measurements 1968-1973. Smithson Radiat Biol Lab. Rockville Md

Klein WH, Goldberg B (1975) Solar radiation measurements 1974/1975. Smithson Radiat Biol Lab, Rockville Md

Knox GA, Lowry JK (1977) A comparison between the benthos of the southern ocean and the northern ocean with special reference to the Amphipoda and the Polychaeta. In: Dunbar (ed) Polar oceans, pp 423-462. Arctic Institute of America, Calgary

Koponen S (1972) On the spiders of the ground layer of a pine forest in Finnish Lapland, with notes on theier diurnal activity. Rep Kevo Subarct Res Stn 9:32-34

Koponen S, Ojala H (1974) On the mesofauna of the field layer of three subarctic habitats. Rep Kevo Subarct Res Stn 11:65-71

Koponen S, Ojala M-L (1975) Quantitative study of invertebrate groups in the soil and ground layer of the IBP sites at Kevo, northern Finland. Rep Kevo Subarct Res Stn 12:45-52

Krüll F (1976a) The position of the sun is a possible zeitgeber for Arctic animals. Oecologia (Berlin) 24:141-148

Krüll F (1976b) Zeitgebers for animals in the continuous daylight of high arctic summer. Oecologia (Berlin) 24:149-157

Krüll F (1976c) The synchronizing effect of slight oscillations of light intensity on activity period of birds. Oecologia 25:301-308

Kureck A (1969) Tagesrhythmen lappländischer Simuliiden (Diptera). Oecologia (Berlin) 2:385-410

Kuty E (1972) Food habits and ecology of wolves on barren-ground caribon range in the Northwest territories. Can Wildl Serv Rep Ser 21:36

Lahti S, Tast J, Uotila H (1976) Fluctuations in small rodent populations in the Kilvisjärvi area in 1950-1975. Luonnon Tutkija 80:97-107

Laine KJ, Niemelä P (1980) The influence of ants on the survival of mountain birches during an Oporinia autumnata outbreak. Oecologia 47

Lasenby DC, Langford RR (1972) Growth, life history and respiration of Mysis relicta in an Arctic and temperate lake. J Fish Board Can 29 (12):1701-1708

Lieth H, Whittaker RH (1975) Primary productivity of the biosphere. Ecol Studies vol XIV 339 pp. Springer, Berlin Heidelberg New York

Likens GE (1975) Primary production on inland aquatic ecosystems. In: Lieth, Whittaker (eds) Primary productivity of the biosphere. Ecol Studies Vol XIV, pp 185-202. Springer, Berlin Heidelberg New York

Lind EA (1974) Fish production in a pond and a lake in northern Finland studied by means of a mark-recapture method, 5 pp. Kalottialueen Rauhanpäivät 5.-7-7-1974, Rovaniemi

Llano A (ed) (1977) Adaptations within Antarcuic ecosystems, 1252 pp. Proc. 3rd SCAR Symp Dallas (Gulf)

Løndevall CF (1952) The bird fauna in the Abisko National Park and its surroundings. K Sven Vetenskapssaad Avhandl I Naturskyddsarden 7:73

Lønø O (1959) Reinen på svalbard. Nor Polarinst Medd 83:3

Lønø O (1960) I. Transplantation of the muskox in europe and North-America. II. Transplantation of hares to svalbard. Nor Polarinst Medd 84:29

Løvenskiold H (1964) Avifauna Svalbardensis. Nor Polarinst Skr 129:7-740

Luick JR, Lent PC, Klein D, White RG (1975) Proceedings of the first international Reindeer and Caribou Symposium, 551 pp. Biol Pap Univ Alaska Fairbanks

Lunde T (1963) Ice conditions at Svalbard 1946 - 1963. Nor Polarinst Arbok

MacLean Jr, SF, Pitelka FA (1971) Seasonal patterns of abundance of tundra arthropods near Barrow. Artic 24:19-40

Madsen H (1940) A study of the littoral fauna of northwest greenland. Medd Grønl 124:4-24

Madsen H (1936) Investigations on the shore fauna of east greenland with a survey of the shores of other arctic regions. Medd Grønl 100:6-79

Manning TH (1971) Geographical variation in the polar bear Ursus maritimus Phipps. Can Wildl Serv Rep 13:27

McAlpine JF (1965) Observations on Anthophilous Diptera at Lake Hazen, Ellesmere Island. Can Field-Nat 79:247-252

McConnaughey T, McRoy CP (1979) Food-web structure and the fractionation of carbon isotopes in the Bering Sea. Mar Biol 53:257-263

McLaren L (1964) Zooplankton of Lake-Hazen, Ellesmere-Island, and a nearby pond, with special reference to the Copepod cyclops scutifer, SARS. Can J Zool 42:613-629

McPherson AH (1965) The origin of diversity in mammals of the Canadian Arctic Tundra. Syst Zool 14:153-173

McPherson AH (1969) The dynamics of the Canadian arctic fox populations. Can Wildl Serv Rep Ser 8:52

Merkel G (1977) The effects of temperature and food quality on the larval development of Gryllus bimaculatus (Orthoptera, Gryllidae). Oecologia (Berlin) 30:129-140

Mieghem J van, van Oye P (eds) (1965) Biogeography and ecology in Antarctica. Monogr Biol 15:762

Miller DR (1976) Biology of the Kaminuriak population of barren-ground caribou, part 3: Taiga winter range relationships and diet. Can Wildl Serv Rep Ser 36:42

Miller F (1974) Biology of the Kaminuriak population of barren-ground caribou. Part 2: Dentition as an indicator of age and sex; composition and socialization of the population. Can Wildl Serv Rep Ser 31:88

Miller F, Russell H, Gunn A (1977) Distributions, movements and numbers of Peary caribou and muskoxen on the western Queen Elizabeth Islands, Northwest Territories, 1972-74. Can Wildl Serv Rep Ser 40:55

Miller FL, Broughton E (1974) Calf mortality on the calving ground of Kaminuriak caribou, during 1970. Can Wildl Serv Rep Ser 26:26

Müller K (1970a) Die Tages- und Jahresperiodik der Buntflossenkoppe Cottus poecilopus Heckel am Polarkreis. Oikos Suppl 13:108-121

Müller K (1970b) Phasenwechsel der lokomotorischen Aktivität bei der Quappe Lota L. Oikos Suppl 13:122-129

Müller K (1973) Circadian rhythms of locomotor activity in aquatic organisms in the subarctic summer. Aquilo Ser Zool 14:1-18

Müller K (1974) Fauna messaurensis. Norrbottens Natur. Norbottens Läns Naturvärdsförbund, Argang. 30, Småskrift 1:1-93

Müller-Haeckel A, Solem JO (1974) Tagesperiodik von Kieselalgen und Grünalgen in einem Gewässer Spitzbergens. Nor Polarinst Arbok S 174-181

Müller-Schwarze D (1965) Zur Tagesperiodik der allgemeinen Aktivität der Weddel-Robbe (Leptonychotes Weddelli) in Hallet, Antarktika, Z Morphol Ökol Tiere 55:796-803

Müller-Schwarze D (1966) Tagesperiodik der Aktivität des Adélie-Pinguins im antarktischen Polartag. Umschau 18:603-604

Murie OJ (1975) A field guide to animal tracks, pp 1-375. Hougton Mifflin Company Boston

Murphy RC (1936) Oceanic Birds of South America. 2 vols, 1245 pp. McMillan, New York

Nettleship DN (1974) The breeding of the knot Calidris canutus at Hazen Camp, Ellesmere Island, N.W.T. Polarforschung 44:8-26

Nettleship ON (1977a) Studies of seabirds at Prince Leopold Island and Vicinity, Northwest territories. Can Wild Serv 73:11

Nettleship D (1977b) Seabird resources of Eastern Canada: Status, problems and prospects. In: Mosquin, Suchal (eds) Canada's species and habitats, pp 96-108. Ottawa

Nettleship DN, Maher WJ (1973) The avifauna of Hazen Camp, Ellesmere Island, N.W.T. Polarforschung 43:66-74

Nettleship DN, Smith PA (eds) (1975) Ecological sites in northern Canada. Can Wildl Serv p 326, Ottawa

Neudecker C (1971) Lokomotorische Aktivität von Carabus glabratus Payk. und Carabus violaceus L am Polarkreis. Oikos 22:128-130

Newsholme EA, Crabtree B, Higgins SJ, Thornton SD, Start C (1972) The activities of fructose diphosphatase in flight muscles from the bumble-bee and the role of this enzyme in heat generation. Biochem J 128:89-97

Nuorteva P (1963) Die influence of Oporinia autumnata (BKh.) (Lep. Geometridae) on the timber-line in subarctic conditions. Ann Entomol Fenn 29:270-277

Nurminen M (1965) Emchytraeid and Lumbricid records (Oligochaeta) from Spitsbergen. Ann Zool Fenn 2:1-17

Nygren J (1978) Interindividual influence of diurnal rhythms of activity in cycling and noncycling populations of the field vole, Microtus agrestis L. Oecologia 35:231-240

Oliver RD (1964) A limnological investigation of a large Arctic lake, Nettilling Lake, Baffin Island. Arctic 17:69-83

Oliver RD (1968) Adaptations of Arctic chironomidae. Ann Zool Fenn 5: 111-118

Oliver RD (1976) Chironomidae (Diptera) of Char Lake, Cornwallis Island, N.W.T., with descriptions of two new species. Can Entomol 108:1053-1064

Oliver RD, Corbet PS (1966) Aquatic habitats in a high arctic locality: the Hazen Camp study area. D Phys R (G) Hazen 26:1-115

Palmgren P (1935) Über den Tagesrhythmus der Vögel im arktischen Sommer. Orn Fenn 12:107-121

Papi F, Syrjämäki J (1963) The sun-orientation rhythm of wolf spiders at different latitudes. Arch Ital Biol 101:59-77

Parker GR (1972a) Biology of the Kaminuriak population of barren-ground caribou. Part 1: Total numbers, mortality, recruitment, and seasonal distribution. Can Wildl Serv Rep Ser 20:95

Parker GR (1972b) Trends in the population of barren-ground caribou of mainland Canada over the last two decades. Can Wildl Serv Occ Pap 10:12

Parker GR (1975) An investigation of caribou range on Southampton Island, Northwest Territories. Can Wildl Serv Rep Ser 33:83

Parker GR (1978) The diets of muskoxen and Peary caribou on some islands in the Canadian High Arctic. Can Wildl Ser Occ Pap 35:21

Parker GR, Ross RK (1976) Summer habitat use by muskoxen (Ovibos moschatus and Peary Caribou (Rangifer tarandus pearyi) in the Canadian High Arctic. Polarforschung 46:12-35

Parmelee DF, Payne RB (1973) On multiple broods and the breeding strategy of Arctic sanderlings. Ibis 114:218-226

Pedersen A (1957) Der Eisbär. Neue Brehm-Bücherei 201, Wittenberg

Pennycuick CJ (1956) Observations on a colony of Brünnich's guillemont Uria lomvia in Spitsbergen. Ibis 98:80-99

Phillipson J (1966) Ecological energetics, 57 pp. Edward Arnold, London

Pohl H, West GC (1973) Daily and seasonal variation in metabolic response to cold during rest and forced exercise in the common redpoll. Comp Biochem Physiol 45 A:851-867

Pruitt WO (1978) Boreal ecology. Inst Biol Stud Biol 91:1-73

Pulliainen, E (1970) Winter nutrition of the rock ptarmigan, Lagopus mutus (Montin), in northern Finnland. Ann Zoll Fenn 7:295-302

Pulliainen E (1971) Nutritive values of some lichens used as food by reindeer in northeastern Lapland. Ann Zool Fenn 8;385-389

Pulliainen E (1973) Winter ecology of the red squirrel (Sciurus vulgaris L.) in northeastern Lapland. Ann Zool Fenn 10:487-494

Pulliainen E, Paloheimo L, Syrjälä L (1968) Digestibility of blueberry stems (Vaccinium Myrtillus) and cowberries (Vaccinium Vitis-Idaea) in the willow grouse (Lagopus Lagopus). Ann Acad Sci Fenn A IV Biol 126:5-14

Rakusa-Suszczewski St (1963) Thermics and chemistry of shallow fresh water pools in Spitsbergen. Pol Arch Hydrobiol 11:169-187

Remane A (1940) Einführung in die zoologische Ökologie der Nord- und Ostsee. In: Grimpe, Wagler (eds) Die Tierwelt der Nord- und Ostsee, Bd 1 S 238. Leipzig

Remmert H (1957) Aves. In: Grimpe and Wagler (eds). Die Tierwelt der Nord- und Ostsee, S 102. Leipzig

Remmert H (1964) Distribution and the ecological factors controlling distribution of the european wrack fauna. Botanica Gothoburgensia II, pp 179-184. Proc 5th Mar Biol Symp, Götheborg

Remmert H (1965a) Biologische Periodik. In: Handbuch der Biologie, Bd V, S 336-410. Akad Verlagsgesellschaft, Frankfurt/Konstanz

Remmert H (1965b) Zur Ökologie der küstennahen Tundra Westspitzbergens. Z Morphol Ökol Tiere 55:142-160

Remmert H (1965c) Über den Tagesrhythmus arktischer Tiere. Z Morphol Ökol Tiere 55:142-160

Remmert H (1968a) Über die Bedeutung volkreicher Meeresvogelkolonien und pflanzenfressender Landtiere für die Tundra Spitzbergens. Veröff Inst Meeresforsch Bremerhaven XI:47-60
Remmert H (1968b) Die Tundra Spitzbergens als terrestrisches Ökosystem. Umsch Wiss Tech 41-44
Remmert H (1976) Gibt es eine tageszeitliche ökologische Nische? Verh Dtsch Zool Ges 1976:29-45
Remmert H (1980) Ecology 1-269. Springer, Berlin Heidelberg NewYork
Remmert H, Wisniewski W (1970) Low resistance to cold of polar animals in summer. Oecologia 4:111-112
Remmert H, Wünderlin K (1970) Temperature differences between Arctic and Alpine meadows and their ecological significance. Oecologia 4:208-210
Richards (1931) Some of the bumblebees allied to Bombus alpinus L. Tromsö Mus Aarskr 50 (1927):1-32
Rivolier J (1957) Gast bei den Kaiserpinguinen, S 9-106. Stuttgart
Rønning OI (1961) The Spitsbergen species of Colpodium trin., Pleuropogon R. Br. and Puccinellia parl. Det KGL Nor Videnskabers Selskabs Skr 4:3-50
Rønning OI (1963) Phytogeographical problems in Svalbard. In: North Atlantic biota and their history, pp 99-107. Pergamon Press, London
Rønning OI (1964) Svalbards flora. Nor Polarinst Polarbok 1:123
Rønning OI (1965) Studies in the Dryadion of Svalbard. Nor Polarinst Skr 134
Rønning OI (1966) Pionerplanter pa veiskraninger i omegnen av Ny-Alesund, Svalbard, Blyttina 24:332-338
Rønning OI (1971) Synopsis of the flora of Svalbard. Nor Polarinst Arbok of 80-93
Rønning OI (1972) The distribution of the vascular Cryptogams and Monocotyledons in Svalbard. Det K Nor Videnskabers Selskab Skr 24
Rosswall T (1971) Systems analysis in Northern coniferous forests – IBP workshop. Bull Ecol Res Comm 14:194
Rosswall T, Heal OW (eds) (1974) Structure and function of tundra ecosystems. Ecol Bull 20:448
Rüppell G (1968) Über Ökologie und Tagesrhythmus von Bodenarthropoden eutrophierter Tundragebiete Westspitzbergens. Pedobiologia 8:150-157
Rüppell G (1969a) Beiträge zum Verhalten des Krabbentauchers (Plautus alle alle). J Ornithol 110:161-169
Rüppell G (1969b) Flugstudien an felsbewohnenden Vögeln (Fulmarus glacialis und Uria lomvia) mit Hilfe kinematographischer Methoden. Res Film 6:445-457
Ryder JP (1967) The breeding biology of Ross' goose in the Perry River Region, NWT. Can Wildl Serv Rep Ser 3:1-56
Salomonsen F (1950-1951) Grönlands Fugle (The birds of Greenland), pp 630. Copenhagen
Salomonsen F (1964) Vogelzug, S 205. BVL Mod Biol Ser, München
Salomonsen F (1967) Migratory movements of the Arctic tern (Sterna paradisaea pontoppidan) in the southern ocean. K Dan Videns Selsk Biol Medd 1-42
El-Sayed SZ, Turner JT (1977) Productivity of the Antartic and tropical subtropical regions: a comparative study. In: Dunbar (ed) Polar oceans. Arct Inst America, Calgary
Scheppenheim, R (1978) Zum Problem des Frostschutzes durch Peptide und Glykoproteide. Diss Kiel S 119
Scherzinger W (1979) Thermospeicherung beim Rauhfußkauz Aegiolus funereus funereus. Anz Orn Ges Bayern 18:184
Schindler DW, Kalff J, Welch HE, Brunskill GJ, Kling H, Kritsch N (1974) Eutrophication in the High Arctic – Meretta Lake, Cornwallis Island (75°N). J Fish Res Board Can 31:647-662
Schmidt WD (1972) Nocturnalism and variance in ambient vapor pressure of water. Physiol Zool 45:302-309
Schramm U (1972) Temperature-food-interaction in herbivorous insects. Oecologia 9:399-402
Shank ChC, Wilkinson PF, Penner DF (1978) Diet of Peary Caribou, Banks Islands, NWT. Arctic 31:125-132
Skogland T (1975) Range use and food selectivity by wild reindeer in southern Norway. In: Luick JR, Lent PC, Klein DR, White RG (eds) Proc 1st Int Reindeer Caribou Symp Fairhanck Alaska, pp 342-358
Shaver GR, Billings WD (1977) Effects of daylength and temperature on root elongation on Tundra graminoids. Oecologia (Berlin) 28:57-65
Skreslet S (1978) Spawning in Chlamis islandica (O.F. Müller) in relation to temperature variations caused by vernal meltwater discharge. Astarte 6:9-14

Smith VR (1976) The effect of burrowing species of procellariidae on the nutrient status of Inland Tussock grasslands on Marion Island. J S Afr Bot 42 (2):265-272

Smith VR (1977) The chemical composition of Marion Island soils, plants and vegetation. S Afr J Antarct Res 7:28-39

Smith RV (1978) Animal-Plant-Soil Nutrient Relationship on Marion Island (Subantarctic). Oecologia (Berlin) 15

Smith ThG (1975) The breeding habitat of the ringed seal (Phoca hispoda). Can J Zool 53:1297-1305

Solhøy T (1976) Terrestrial Gastropods (Mollusca, Gastropoda: Basommatophora and Stylommatophora). Zool Mus Univ Bergen. Fauna Hardangervidda 10:24-45

Steigen A (1973) Sampling invertebrates active below a snow cover. Oikos 24:373-376

Steigen A, Solhøy T, Gyllenberg G (1975) Energy budget of a population of adult Carabodes labyrinthicus (Mich.) (Acari, Oribatei). Norw J Entomol 59-61

Stirling I (1974) Ecology of the weddell seal in McMurdo sound Antarctica. Ecology 50:573-586

Stirling I, Jonkel Ch, Smith P, Robertson R, Bross D (1977) The ecology of the polar bear (Ursus maritimus along the western coast of Hudson Bay. Can Wildl Serv Occ Pap 33:64

Stonehouse B (1972) Tiere der Antarktis, S 172. BLV, München Bern Wien

Strand A (1972) Die Käferfauna von Svalbard. S Nor Entomol Tidsskr VI:2-17

Stross RG (1969) Photoperiod control of diapause in Daphnia. III: Twostimulus control of long-day, short-day induction. Biol Bull 137:359-374

Stross RG, Kangas DA (1974) The reproductive cycle of Daphnia in an arctic pool. Ecology

Sumerhayes VS, Elton Ch (1923) Contributions to the ecology of Spitsbergen and Bear Island. J Ecol 11:214-286

Sumerhayes VS, Elton Ch (1928) Further Contributions to the Ecology of Spitsbergen. J Ecol 16:193-268

Svenden P (1959) The algal vegetation of Spitsbergen. Nor Polarinst Skr 116:5-47

Syrjämäki J (1968) Diel patterns of swarming and other avtivities of two arctic Dipterans (Chironomidae and Trichoceridae) on Spitsbergen. Oikos 19:250-258

Tanhuanpää E, Pulliainen E (1969) Molar fatty acid composition of some organ fats in the willow grouse (Lagopus Lagopus) and the rock ptarmigan (Lagopus Mutus). Ann Acad Sci Fenn A IV Biol 141:4-14

Tast J, Kalela O (1971) Comparisons between rodent cycles and plant production in Finnish Lapland. Ann Acad Sci Fenn A IV Biol 186:1-14

Teckelmann U (1974) Temperaturwirkungen auf Wachstum und Stoffwechsel kaltstenothermer Fließwassertiere. Arch Hydrobiol 74:479-527

Tener J (1972) Queen Elizabeth Islands game survey, 1961. Can Wildl Ser Occ Pap 4:50

Theede H (1967) Probleme der Frostresistenz bei Meerestieren. Naturwiss Rundsch Sonderdr 20:468-475

Thiele H-U (1976) Tageslängenmessung als Grundlage der Jahresrhythmik des Laufkäfers (Pterostichus nigrita F.). Verh Dtsch Zool Ges

Thienemann A (1950) Verbreitungsgeschichte der Süßwassertierwelt Europas. Die Binnengewässer 18:809 pp. Stuttgart

Thomas DC (1969) Population estimates and distribution of barren-ground caribou in Mackenzie District, NWT, Saskatchewan, and Alberta-March to May 1967. Can Wildl Serv Rep Ser 9:44

Thorsteinsson I, Gardasson A, Olafsson G, Gudbergson GM (1970) Islenzu hreindýrin og sumarlönd peirra. Náttúrufraedingurinn 40:145-170

Tieszen LL (ed) (1978) Vegetation and production ecology of an Alaskan Arctic tundra. Ecol Studies vol 29, pp 686. Springer, Berlin Heidelberg New York

Tilbrook PJ (1970) The Terrestrial environment and invertebrate fauna of the maritime Antarctic, Antarctic. Ecology 2:886-896

Troll C (1955) Der jahreszeitliche Ablauf des Naturgeschehens in den verschiedenen Klimagürteln der Erde. Stud Gen 12:714-732

Vaartaja O (1959) Evidence for photoperiodic ecotypes in trees. Ecol Monogr 29:91-111

Väisenen RA, Järvinen O (1977) Structure and fluctuation of the breeding bird fauna of a north Finnish peatland area. Ornis Fenn 54:143-153

Vikberg V (1973) 17. List of the hymenoptera collected by the finnish Spitsbergen expeditions. Ann Entomol Fenn 39:67-68

Vockeroth JR (1956) Distribution pattern of the Scatomyzinae (Diptere, Muscidae). Proc 10th Int Congr Entomol 1:619-626
Voous KH (1964) Die Vogelwelt Europas und ihre Verbreitung, 59-284. Hamburg
Wagner G (1958) Beobachtungen über Fütterungsrhythmus und Nestlingsentwicklung bei Singvögeln im arktischen Sommer. Ornithol Beob Heft 2:37-54
Wallgren H (1954) Energy metabolism of two species of the genus emberiza as correlated with distribution and migration. Acta Zool Fenn 84:5-110
Watermann TH (ed) (1965) The physiology of Crustacea, 2 vol, 432, 429 pp. Academic Press, London New York
Watson A (ed) (1970) Animal Populations in Relation to their food resources, 477 pp. Oxford & Edinburgh
Weiss W (1975) Arktis, 188 pp. Wien München
West G (1968) Bioenergetics of captive willow ptarmigan under natural conditions. Ecology 49:1035-1045
West GC (1972) Seasonal difference in resting metabolic rate of Alaskan Ptarmigan. Comp Biochem Physiol 42 A:867-876
West GC, Pohl H (1973) Effect of the light-dark cycles with different LD-time ratios and different LD intensity-ratios on the activity rhythm of chaffinches. J Comp Physiol 83:289-302
West GC, Salo AC (1979) Seasonal changes in the water and fatty acid composition of the catkin buds of the Alaska Willow. Oecologia 41:207-218
West GG, DeWolfe B (1974) Populations and energetics of Taiga Birds near Fairbanks alaska. Auk 91:757-775
Wielgolaski FE (ed) (1975) Fennoscandian tundra ecosystems. Ecol Studies, vol 16, 366 pp, vol 17, 326 pp. Springer, Berlin Heidelberg New York
Wielgolaski FE, Rosswall Th (eds) (1972) Tundra Biome, 320 pp. IBP, IV Meet, Stockholm
Wieser W (ed) (1973) Effects of temperature on ecothermic organisms, 298 pp. Springer, Berlin Heidelberg New York
Willgohs JF (1961) The white-tailed eagle Haliaetus albicilla (Linné) in Norway. Arb Univ Bergen Math Naturw Ser 12:10-212
Williams N The history of the introduced Reindeer of South Georgia, Deer. 4: 5, 256-261
Winter JE (1969) Über den Einfluß der Nahrungskonzentration und anderer Faktoren auf Filtrierleistung und Nahrungsnutzung der Muscheln Arctica islandica und Modiolus modiolus. Mar Biol 4:88-135
Zenkewitch L (1963) Biology of the seas of the USSR, 995 pp. Allen and Unwin, London
Zhadin VI, Gerd SV (1963) Fauna and flora of the rivers, lakes and reservoirs of the USSR. (Translated from Russian, Israel Program for Scientific Translations, Jerusalem, 626 pp)

Subject Index

adder 34, 164
Adelie penguin, see penguins
Aedes 64
Aegiolus 51, 71
agriculture 6, 161
Alcidae 10, 20, 21, 41, 46, 61, 198
Algae 57
Allen's law 66
Alopex, see fox
Anatidae 10
Andes 3
annual cycle: insects 101, 102
antarctic 1, 94, 119, 225
Anthus 20, 231
antifreeze 35, 47, 53
antlers 94
Apodemus 74
Aptenodytes 46, 48, 49, 94, 234
arctic basin: temperature 218
– – topography 217
– circle 7
– hare, see also hare 134, 181
Ascophyllum 144
Astarte 147
auks, see also Alcidae 10, 38, 61, 198
azimuth position 18

Balanus 144, 218
basal metabolic rate (BMR) 100, 126
beavers 10
„Beharrungstendenz" 82
Belgica 232
Benthos 147, 151, 211, 212, 230
Bergmann's law 67
Bering Sea food web 210
Betula, see birch
biomass 107f.
bipolar distribution 119
birch 1, 45, 131, 161
bird colonies 104, 189, 204, 205
– rocks 59, 104, 198, 216
bird-song 60

body temperature 66
Boloria 34, 35
Bombus, see bumble-bee
Boreus 42
Bovallia 30
Brachionus 33
Braconids 101
Branchinecta 137
bryosystem 233
Bryozoa 211
bumble-bee 34–36, 136, 166, 183
burbot 22, 141
butterflies 33, 136, 176, 183, 184, 194

caddies fly 173, 184
Calanids 219, 222, 223
Calcium 149
Calidris 41, 60, 68, 69, 85
camel 57, 58
Canis, see wolf
capercallie 71, 97
Capreolus, see alo roedeer 97
carabid 23, 28
Carduelis 7, 15, 20, 22, 46
caribou, see reindeer
Cassiope 45, 190
castor 10
catabolism 107
Cephalopods 227, 230
Cetraria 79
chaffinch 10
chalikosystem 233
char, see also Salvelinus 30
Chionea 42
Chionis 231, 232
chironomids 59, 60, 108, 150
chlamydomanas 41
Chlamys 58
Cicadas 102
Citellus 181
Cladocera, see also Daphnia 234
Cladonia 79
Clangula 41, 211
climate and diversity 159
Clunio 144

cockroaches 33
cod 147
Collembola 29, 131, 232
Colobanthus 232
colour temperature 15, 16, 18, 108
competition 64, 202
constancy 129
continental climate 156f.
continuous darkness 20
Copepods 208, 234
coral reef 86, 129
cormorants 21
cosine law 15
Cottus (see also Myxocephalus) 11, 141, 152
crickets 33
Crocethia 68
Cryptopygus 232
Culicids 64, 203
cycles 70, 72, 73, 78, 176
Cygnus 189

Daphnia 19, 23, 102, 208
darkness 20, 21, 64
deciduous trees 160
defaecation 51
Deschampsia 232
Dicrostonyx, see Lemming
diurnal rhythmicity 7
diver 88
diversity 120, 159, 175, 221
Dolichopodids 102, 176
domestication 63, 174
Dromas 86
Drosophila 31
Dryas 60
ducks 10, 209, 211
dung 134, 231, 200f.
dunlin 68

earthworms 135
ecological efficiency 99, 123
– laws 66
egg-harvest 209
eggs 38
eider duck 38
Emberiza 21, 66

Empidids 102, 196
endothermy 109
energy 25
– flow 131, 184, 185, 186, 187, 188, 229
– –, flow, Antarctic 153
Ephemeroptera 211
epifauna 148
Epistrophe 102
ermine 183
Eskimo dog 52
Euphausiaceae, see also Krill 30, 222, 223
eutrophication 104, 105, 106, 141, 231
Exechia 64
exposition 195
eyes 51

faeces 127, 198, 200
Falco 21, 81, 111
farming 6, 161
fat 33
faunal lists 159, 167ff., 177ff., 186, 187, 197, 211–213, 229, 230, 233
Finland 103
fish 123, 206, 209, 211, 213, 219, 222, 223, 224
flying squirrel 7, 22
fog 157
food chain 230
– consumption 98, 99, 100, 122, 124, 125, 126
– –, young animals 123 f.
– cycle 224, 225
– supplies 76
– web: Barents sea 220, 222
forest reindeer 174
fox 19, 71, 99, 170, 183
frog 164
frost 35
fulmar 39, 97, 98, 119, 232

Gadus, see also Cod 53
Gammarus 218
Gasterosteus 209
Gavia 84, 88
geese 93, 123 f., 132, 186, 189, 203, 204 f.
geese colonies 129, 189, 204 f.
glacial refugia 111–113, 116, 117, 120
glaciation 110
glacier fauna 42
Glaucomys 7, 22

glutton 59, 96, 170
Glycerine 35, 53
Gold hammer 65, 66
golden hamster 7
– plover 89, 135
great auk 64
greenfinch 7, 10, 15
grit 44, 92
ground squirrel 55, 181
grouse, see also Lagopus, Ptarmigan 22, 42, 59, 71
growing season 1
growth 32, 33, 58, 122, 230
Gryllus 34
guillemot 38, 39, 60 f., 83, 198
Gulf of Bothnia 147, 150
gulls 113, 117
Gulo 59, 96, 170
Gynaephora 184

halibut 147
Halicryptus 150
Hardangervidda 157, 166 ff.
hare 59, 112, 184
hazel hen 43
herbivores 32, 59
– impact of 79, 80
herring 222, 223
Hesse's Law 66
hibernators 22, 55–57
honey bee 166
hormone 15
hot springs 207
humming bird 57
hybridization 119
Hymenoptera 59

ice algae, see also glacier fauna 42, 210
Iceland 6, 59, 74, 93, 120, 203
Ichneumonids 101
imprinting 64
infauna 148
insect, hervivores 131
insolation 25, 47
internal clock 4
introduction of species 120
invasions 81
Irkutsk 5
irradiation 34
islands 103
isotherm 4
ivory gull 39, 47, 82, 165

Jaeger, see skua; stercovarius
Jan Mayen 157

Kilimanjaro 3
kittiwake 106
knot 60
Kongsfjord 104
krill, see also Euphausia 30, 153, 226

Lacerta 34, 164
Lagopus, see Grouse, Ptarmigan 21, 22, 41, 42, 59, 76
lakes 57, 108, 137 ff., 203 ff.
Laphygma 33
Laps 174
Larus, see Gulls
ledipurus 138, 209
lemming, impact of 133, 135
lemmings 41, 42, 52, 64, 68, 70, 73, 77, 88, 95, 112, 176, 183
lichens 176, 202, 203
light intensity 12, 13
Litorina 144, 218
lizard 34, 164
longwave radiation 16
Longyearbyen 1, 15
loon 84, 88
Lota 22, 141
lunar rhythm 24
Lymnae 209, 211
lynx 72 ff.

Macoma 30, 144, 147
– -community 148
magpie 38
mammoth 58
Manatee 126
Mantis 57
marine benthos 144 ff., 147
– faunal elements 57, 119, 204
– ice conditions 142, 145, 146
– relics 57, 119, 204
Marion Island 104
Mc. Murdo 4
mean temperatures 1
Mesidothea 119, 147, 150, 204
Mesocricetus 7, 12
Messaure 18, 166 ff.
Microtus 22, 41, 70
Migrations 64, 82
Monodon 92
moonlight 144
mortality 68
mosquitoes 64, 181, 182
mosses 10, 234
Motacilla 20
moths 29, 136, 184
moulting problems 166

Subject Index

Mount Kenya 3
mountain streams 140
muskox 21, 29, 60, 88, 116, 158, 180, 181
Musteal 21, 42
Mya 144, 147
mycetophilids 59, 65
Mysis 29, 119, 204
Myvatn 161
Myxocephalus, see also Cottus 53

nectar 38
niche 60, 86
– breadth 59
nidifugous birds 68
night 20, 21, 46, 223
nightingale 109
Non-passerines 10
Norway 72, 79, 81
Nyctaea, see snowy owl

Oceanic 59
– climate 156f.
– currents 143
Oenanthe 20, 38, 87
Operophtera 164
opisthobranchs 148
Oporinia 132, 164
ortolan 21, 65
„Ortstreue" 80
Ostracoda 234
overfishing 130
overwinter 55
ovibos, see muskox
owls 51
Oxford 6
Oxygen consumption 26

Pagodroma 39, 232
Pagophila 39, 47, 82, 165
Panolis 31
parasites 194, 233
Parochlus 232
Parus 45, 66
passerines 175
Peary reindeer 203
penguins 46, 94, 231
permafrost 2, 54, 55, 130
Persian Gulf 107
Phalacrocorax 21, 232
Phalarope 86
Phoca hispida 43, 51, 119, 204
Phormia 33
phosphorus 104, 135
photoperoid 4, 23, 102
photosynthesis 19, 26, 58, 107, 138

physiological clock 9
phytochrome 16
Pica 38
Picea 1
pine forest 1, 131
pinguinus 64
Pinus 1, 131
pipeline 176
pipit 20, 231
placenta 231
plankton 19, 208, 219, 221, 222
– larvae 149
– production 19, 206–208, 214
– vertical migrations 18f., 206
plecoptera 140
Plectrophenax, see snow bunting 15, 19, 38
Pleistocene 58, 110
Plutella 194
Pluvialis 87, 88, 89
polar bear 21, 43, 44, 46, 62, 118
– light 20
pollination 135, 195, 196
ponds 140
Pontoporeia 119, 150, 204
population density 77, 127, 128, 177, 182, 185, 186
potassium 104
precipitation 156
precision (rhythm) 13
predator, effect of 78, 135
predators 64, 77, 80, 123, 181, 186
primary production 19, 158, 200, 203, 206, 207, 213, 226, 230
productivity 26, 107, 109
–, tundra 134
Pseudalibrothus 153, 218
Ptarmigan see also Grouse, Lagopus 21, 59, 74, 88
Pterostichus 23, 28, 29
puffin 61
Pungitius 141, 173, 211, 213

Q_{10} 26
Quito 5

radiation 4, 25, 57, 59, 156f.
rainfall 59, 161
Rana 164
Rangifer, see reindeer
Ranunculus 8
razorbill 61–63
red deer 92

redpolls 20, 22
redwing 38
reindeer (=caribou) 19, 21, 41, 52, 60, 64, 79, 90, 94, 110, 114–116, 127, 176, 180, 183, 199
–, age-structure 201
–, fjell 111, 115, 174
–, food composition 176
–, forest 111, 115, 174
–, impact on system 202
–, mortality 201
–, wild 174
Reproductive cycle 94–96
reptiles 32
resilience 129, 130
respiration 51
ressources, Antarctica 227
Rhamphomyia 102, 196
Rhodostethia 47, 82
Rhynchaenus 45, 194
rhythm 68
ringed seal 43, 51, 119, 204
river climate 57
roads 130, 176
roe deer 97
roseate gull 47
rotifers 33, 57, 208
Rubus 45, 196
running water 55

salinity 58
Salix 38, 45, 184
Salmo 141, 209
Salvelinus 30, 141, 209, 213
sanderling 68
sandpiper 41
sawflies 194
Saxifraga 102
Scandinavia 81
sciarids 59, 64
Scoresby Sound 107
seacow 126
seagulls, see also gulls 10
sealeopard 234
seals 19, 44, 62, 92, 95, 96, 154, 227, 228, 231, 234
seasonal rhythm 219, 220, 222–224
seed production 76
sheath bill 231, 232
Simuliidae 18, 67, 203
siskin 10
skua 84, 119, 231
sleeping communities 46
snow 42
– bunting 9, 15, 19, 60, 69, 87

snow hollows   44
– petrel   39
snowshoe hare   57
snowy owl   7, 21, 63, 77, 98, 100, 123 f., 125
soil organisms   178, 185, 186, 194
solar constant   25
– energy   57, 234
– –, traps   57, 139
– radiation   4, 59, 156 f.
Somateria   38
South Georgia   120
spectral composition   15, 17, 18, 103
Spermophilus   55
Sphaeriidae   108, 141
Spitsbergen   7, 32, 75, 102, 158
spruce   1, 76
squids   227, 230
stability   129
standing stock   107
starling   17, 20
Stercorarius   39, 41, 71, 84
Sterna   19, 39, 84, 233
stress   77
Sturnus   17, 20
sunshine   67, 195
swan   189
swim-bladder   138

synchronization   13
Syrphids   102

Taraxacum   8
teleosts   54
temperature   13, 17, 24 ff., 94, 182
– fluctuations   31, 139, 155
– gradient   39
Tenthredinids   101
terns   10, 19, 84, 85, 233
territories   67, 68, 232
Tetrao   97
Thalarctos, see polar bear
Thermoisopleth   5
tidal regions   54
timberline   64, 133, 161
tipulids   131, 183
tit   45, 66
trees: annual rings   132
trichocerids   59
Trichoptera   173, 184
Trifolium   8
tropical rainforest   129
tropics   86
Tularemia   71
Tundra   1, 157
Tupaia   7, 12
Turdus   38

unicentric species   165
unpredictability   77, 160
Uria, see Guillemot   39, 83, 198
uric acid   77
urination   51

Vikings   130, 174
Vipera   34, 164
voles   22, 70

wagtail   20
wasps   166
weasel   21, 42, 43, 81
Weber-Fechner law   20
Weddell seal   19
weight, annual cycle   97
whale   91, 95, 97, 100, 226–228
– hunting   130, 190, 227 ff.
wheatear   20, 38, 87
wind speed   50, 57, 182
winglessness   67
winter   21, 22, 43–45, 47, 52, 53, 55, 223
Wisconsin Glaciation   111
wolf   21, 29, 113, 183
Wolverine, see Gulo, Glutton
woolly rhinoceros   58

Zeitgeber   4 ff., 58

H. Remmert
Ecology
A Textbook

Translated from the German by
M. Biederman-Thorson

1980. 189 figures, 12 tables. Approx. 300 pages
ISBN 3-540-10059-8

The second edition of this outstanding textbook is now available in translation to English-speaking readers. Revised and expanded from the first edition, it brings into even greater focus the relationship between ecology and sensory physiology.

From the reviews of the first German edition:
"The literature is not exactly poor in attempts to describe ecological relationships in textbook form. What makes Remmert' book stand out from the rest is the author's dynamic perspective, his simple, flowing style, and, in many instances, the interpretations themselves... Appropriately enough, the coevolution of the various members of an ecosystem is given special emphasis. This point usually receives inadequate attention in comparable textbooks. The author tackles generally accepted or postulated trends, discusses them, reformulates them, and, wherever possible, authenticates and explains them with individual analyses and case studies..."
translated from: *Helgoländer wissenschaftl. Meeresuntersuchungen*

W. Larcher
Physiological Plant Ecology

Translated from the German by
M. A. Biederman-Thorson

2nd totally revised edition. 1980. 193 figures, 47 tables. XVII, 303 pages
ISBN 3-540-09795-3

The rapid advances made in the fields of plant physiology and ecology since this book's first edition have made complete revision of its material necessary. The edition presents information and examples from all climatic zones according to the latest available data. The literature cited now comprises almost 800 references.
This up-to-date introductory textbook with its wealth of data and examples is a must for every ecophysiology student.

From the reviews:
"Dr. Biederman-Thorson's translation of... is most welcome as it ensures a wider audience for this excellent eco-physiological text... The book is almost certainly the best comprehensive text on physiological plant ecology that is currently available and merits a place in every botanical library."
The Journal of Ecology

L. Irving
Arctic Life of Birds and Mammals
Including Man

1972. 59 figures. XI, 192 pages
(Zoophysiology and Ecology, Volume 2)
ISBN 3-540-05801-X

"This book offers very satisfying reading. It is rich with the findings of a senior and respected naturalist and physiologist who has specialized in the northern homotherm fauna, and it deals well with the literature in general...
The chapters on metabolic rates, thermal adaptation and insulation, are excellent and need very little comment by your reviewer...
this book maintains a high standard, and there are some very bright lights in it..."
Polar Record

Springer-Verlag
Berlin
Heidelberg
New York

Ecological Studies

Analysis and Synthesis

Editors: W. D. Billings, F. Golley
O. L. Lange, J. S. Olson

A Selection

Volume 29

Vegetation and Production Ecology of an Alaskan Arctic Tundra

Editor: L. L. TIESZEN
1978. 217 figures, 115 tables. XVII, 686 pages
ISBN 3-540-90325-9

The discovery of large oil deposits on the Alaskan Arctic Coastal Plain took place during a period of expansion in the world-wide awareness of environmental matters. The period 1970-74 was particularly interesting, since the environmental concerns of private industry, the federal government, and the public, were brought to a common focus. In the Alaskan Artic, that focus was the U. S. Tundra Biome Program, and this volume represents the fruit of the intensive and integrated botanical field research of those four years.
In addition to summarizing the most significant results and implications of the program, the volume also provides an excellent general introduction to the environment and ecology of the Coastal Plain Tundra. Following a brief overview of numerous ecosystem components, the main body of the text features.
- a thorough description of vegetation components
- a floristic analysis of mosses and vascular plants
- detailed production and physiological studies
- analyses of photosynthesis, root growth, ecosystem CO_2
- exchange, an water relations.

Vegetation and Production Ecology of an Alaskan Arctic Tundra will provide a comprehensive background for students and researchers interested in tundra and arctic ecology. It may, furthermore, facilitate the resolution of future environmental issues.

Volume 32

Perspectives in Grassland Ecology

Results and Applications of the US/IBP Grassland Biome Study

Editor: N. R. FRENCH
1979. 60 figures, 47 tables. XI, 204 pages
ISBN 3-540-90384-4

Perspectives in Grassland Ecology examines the biomass structure of grasslands in North America and other areas of the world. It compares trophic structure, and describes structure modifications occuring when water and nutrients are artificially supplied. The most important driving variables, processes, and the major functional aspects of production are analyzed and modeled mathematically. Unique in its detailed comparative analysis of the major grassland types, this book will be appreciated by researchers in range science, ecology, agronomy and botany.

Volume 16

Fennoscandian Tundra Ecosystems
Part 1: Plants and Microorganisms

Editor: F. E. WIELGOLASKI
Editorial Board: P. KALLIO, T. ROSSWALL
1975. 90 figures, 96 tables. XV, 366 pages
ISBN 3-540-07218-7

Volume 17

Fennoscandian Tundra Ecosystems
Part 2: Animals and Systems Analysis

Editor: F. E. WIELGOLASKI
Editorial Board: P. KALLIO, H. KAURI, E. ØSTBYE, T. ROSSWALL
1975. 81 figures, 97 tables. XIII, 337 pages
ISBN 3-540-07551-8

"...The present two volumes which report on the findings and their implications of the tundra ecosystems projects in Finland and the Scandinavian peninsula are of the latter category. In roughly 700 pages they give a most valuable account of the complexity of factors and organisms constituting the tundra ecosystems, and the functioning of such ecosystems. ...the editor provides a summary and conclusion of the data presented in both volumes. It is rejoicing and important that the results of these accomplished studies collectively have been made available to the international scientific community in an easily attainable form. The editor and members of the editorial board should be congratulated with their achievement." *Vegetatio*

Springer-Verlag
Berlin Heidelberg NewYork